Global Warming Trends

GL**O**BAL
WARMING

Global Warming Trends

Ecological Footprints

Julie Kerr Casper, Ph.D.

Facts On File
An imprint of Infobase Publishing

GLOBAL WARMING TRENDS: Ecological Footprints

Facts On File, Inc.
An imprint of Infobase Publishing
132 West 31st Street
New York NY 10001

Library of Congress Cataloging-in-Publication Data

Casper, Julie Kerr.
  Global warming trends : ecological footprints / Julie Kerr Casper.
      p. cm.— (Global warming)
  Includes bibliographical references and index.
  ISBN-13: 978-0-8160-7261-3
  ISBN-10: 0-8160-7261-2
  1. Paleoclimatology. 2. Global warming. I. Title.
  QC884.C37 2009
  551.6—dc22   2008046358

Text design by Erik Lindstrom
Illustrations by Melissa Ericksen and Sholto Ainslie
Photo research by the author

Printed in the United States of America

Bang Hermitage 10 9 8 7 6 5 4 3 2 1

This book is printed on acid-free paper.

CONTENTS

PREFACE

We do not inherit the Earth from our ancestors—
we borrow it from our children.

This ancient Native American proverb and what it implies resonates today as it has become increasingly obvious that people's actions and interactions with the environment affect not only living conditions now, but also those of many generations to follow. Humans must address the effect they have on the Earth's climate and how their choices today will have an impact on future generations.

Many years ago, Mark Twain joked that "Everyone talks about the weather, but no one does anything about it." That is not true anymore. Humans are changing the world's climate and with it the local, regional, and global weather. Scientists tell us that "climate is what we expect, and weather is what we get." Climate change occurs when that average weather shifts over the long term in a specific location, a region, or the entire planet.

Global warming and climate change are urgent topics. They are discussed on the news, in conversations, and are even the subjects of horror movies. How much is fact? What does global warming mean to individuals? What should it mean?

The readers of this multivolume set—most of whom are today's middle and high school students—will be tomorrow's leaders and scientists. Global warming and its threats are real. As scientists unlock the mysteries of the past and analyze today's activities, they warn that future generations may be in jeopardy. There is now overwhelming evidence that human activities are changing the world's climate. For thousands of years, the Earth's atmosphere has changed very little; but today, there are problems in keeping the balance. Greenhouse gases are being added to the atmosphere at an alarming rate. Since the Industrial Revolution (late 18th, early 19th centuries), human activities from transportation, agriculture, fossil fuels, waste disposal and treatment, deforestation, power stations, land use, biomass burning, and industrial processes, among other things, have added to the concentrations of greenhouse gases.

These activities are changing the atmosphere more rapidly than humans have ever experienced before. Some people think that warming the Earth's atmosphere by a few degrees is harmless and could have no effect on them; but global warming is more than just a warming—or cooling—trend. Global warming could have far-reaching and unpredictable environmental, social, and economic consequences. The following demonstrates what a few degrees' change in the temperature can do.

The Earth experienced an ice age 13,000 years ago. Global temperatures then warmed up 8.3°F (5°C) and melted the vast ice sheets that covered much of the North American continent. Scientists today predict that average temperatures could rise 11.7°F (7°C) during this century alone. What will happen to the remaining glaciers and ice caps?

If the temperatures rise as leading scientists have predicted, less freshwater will be available—and already one-third of the world's population (about 2 billion people) suffer from a shortage of water. Lack of water will keep farmers from growing food. It will also permanently destroy sensitive fish and wildlife habitat. As the ocean levels rise, coastal lands and islands will be flooded and destroyed. Heat waves could kill tens of thousands of people. With warmer temperatures, outbreaks of diseases will spread and intensify. Plant pollen mold spores in the air will increase, affecting those with allergies. An increase in severe weather could result in hurricanes similar or even stronger than Katrina in 2005, which destroyed large areas of the southeastern United States.

Higher temperatures will cause other areas to dry out and become tinder for larger and more devastating wildfires that threaten forests, wildlife, and homes. If drought destroys the rain forests, the Earth's delicate oxygen and carbon balances will be harmed, affecting the water, air, vegetation, and all life.

Although the United States has been one of the largest contributors to global warming, it ranks far below countries and regions—such as Canada, Australia, and western Europe—in taking steps to fix the damage that has been done. Global Warming is a multivolume set that explores the concept that each person is a member of a global family who shares responsibility for fixing this problem. In fact, the only way to fix it is to work together toward a common goal. This seven-volume set covers all of the important climatic issues that need to be addressed in order to understand the problem, allowing the reader to build a solid foundation of knowledge and to use the information to help solve the critical issues in effective ways. The set includes the following volumes:

Climate Systems
Global Warming Trends
Global Warming Cycles
Changing Ecosystems
Greenhouse Gases
Fossil Fuels and Pollution
Climate Management

These volumes explore a multitude of topics—how climates change, learning from past ice ages, natural factors that trigger global warming on Earth, whether the Earth can expect another ice age in the future, how the Earth's climate is changing now, emergency preparedness in severe weather, projections for the future, and why climate affects everything people do from growing food, to heating homes, to using the Earth's natural resources, to new scientific discoveries. They look at the impact that rising sea levels will have on islands and other areas worldwide, how individual ecosystems will be affected, what humans will lose if rain forests are destroyed, how industrialization and pollution puts peoples' lives at risk, and the benefits of developing environmentally friendly energy resources.

The set also examines the exciting technology of computer modeling and how it has unlocked mysteries about past climate change and global warming and how it can predict the local, regional, and global climates of the future—the very things leaders of tomorrow need to know *today*.

> *We will know only what we are taught;*
> *We will be taught only what others deem is important to know;*
> *And we will learn to value that which is important.*
> —Native American proverb

ACKNOWLEDGMENTS

Global warming may very well be one of the most important issues you will have to make a decision on in your lifetime. The decisions you make on energy sources and daily conservation practices will determine not only the quality of your life, but also that of your future descendents.

I cannot stress enough how important it is to gain a good understanding of global warming: what it is, why it is happening, how it can be slowed down; why everybody is contributing to the problem; and why *everybody* needs to be an active part of the solution.

I would sincerely like to thank several of the federal government agencies that research, educate, and actively take part in dealing with the global warming issue—in particular, the National Aeronautics and Space Administration (NASA), the National Oceanic and Atmospheric Administration (NOAA), the Environmental Protection Agency (EPA), and the U.S. Geological Survey (USGS)—for providing an abundance of resources and outreach programs on this important subject. I give special thanks to Al Gore and Arnold Schwarzenegger for their diligent efforts toward bringing the global warming issue so powerfully to the public's attention. I would also like to acknowledge and give thanks to the many wonderful universities across the United States and in England, Canada, and Australia, as well as private organizations, such as the World Wildlife Fund, that diligently strive to educate others and help toward finding a solution to this very real problem.

I want to give a huge thanks to my agent, Jodie Rhodes, for her assistance, guidance, and efforts; and also to Frank K. Darmstadt, my editor, for all his hard work, dedication, support, and helpful advice and attention to detail. His efforts to bring this project to life were invaluable. Thanks also to the copyediting department for their assistance and the outstanding quality of their work, with a special thanks to Alexandra Lo Re for her enthusiasm and input.

INTRODUCTION

The Earth's climate is always changing. The most obvious changes that have taken place over the course of geologic time are the shifts from glacial (ice age) periods to interglacial (non–ice age) periods. The Earth has had several "ice ages" throughout time, during which ice covered large portions of the Earth's surface, then melted back to nearly nothing. Scientists know the Earth's climate is always naturally changing, but with the amount of human interference today, the questions have become: How are humans changing it? Are they endangering the future? And what will the Earth be like for our children, grandchildren, and future generations? These are the very issues that I would like you—the reader—to ponder as you gain insight into the problem. What long-term effects will our current decisions and lifestyle choices have on future generations? How much hardship are we adding in and above the natural fluctuations in climate that have been normal for the Earth throughout geologic time? For better or worse, our actions today *will* affect future generations and their standard of living. That is a serious responsibility—not one to be taken lightly.

Climate scientists have repeatedly told us that the 1990s was the warmest decade ever recorded since temperatures began to be consistently kept in the mid-1800s. Scientists around the world support this notion; there are an abundance of data collected worldwide that prove this. Some may argue that the trend has not been a steady upward climb, but that it has had intermittent cooling periods. But, unfortunately, it does not let us off the hook. Although these intervals have occurred every few decades in some locations, the overall trend has been a steady upward climb, coinciding with melting glaciers, rising sea levels, shifting climatic zones, and changing ecosystems worldwide.

Not only were all 10 years of the 1990s among the 15 warmest ever recorded, but also six of them have the distinction of being the *warmest years on historical record—ever.* When scientists look at this trend, what it tells them right away is that the Earth's atmosphere has become significantly warmer over the past 150 years. This notion then raises other questions, such as the following:

- Has this happened before?
- What effect did warmer temperatures have on the atmosphere?
- What effect did warmer temperatures have on plants and animals?
- How have warmer climates affected humans?
- What do warmer climates do to ecosystems?
- If temperature fluctuations occur, how often and how long do they typically last?
- Did animal species in the past migrate, adapt, or go extinct?
- What can we really expect to happen today?
- Can technology solve all our problems?

These are the questions this volume will address. If scientists can obtain a good understanding of past climate change and how susceptible or resilient the environment has been in response, it offers them the valuable insight they need to understand the global changes happening today in the climate.

One way that scientists are able to study global warming is if they have a baseline to compare it to. This baseline is the Earth's past climate. If scientists can successfully reconstruct the Earth's past climate and determine (1) how often it warmed up or cooled down, (2) to what magnitude it warmed up/cooled down, and (3) what mechanisms triggered the warming/cooling phases, then they are in a much better position to understand what global changes they are seeing today and what they mean for humans and the environment—topics you will be exploring in this volume.

Climatologists have written climate records going back only to the mid-1800s. Luckily for them, the Earth has left a long and detailed record of the past. When scientists study ancient climates, a field called paleoclimatology, they are able to reconstruct the Earth's climate back over long periods of time. By looking backward in time, scientists can find evidence of natural processes that may be currently in effect, causing the global temperature changes we are seeing today. When scientists study warm periods of the past, they can discern vital clues that explain the warming of today and whether this is from natural processes or something else, such as something humans have directly caused.

The goal of this volume is to show exactly how scientists determined what the climate was like at a specific time period and how they inferred what global warming may have been like at another time in the past. For instance, if global temperatures rise, then carbon dioxide (CO_2) is released from the ocean. If CO_2 is released, it magnifies the greenhouse effect, making atmospheric temperatures warmer. Conversely, when temperatures drop, then CO_2 enters the oceans, and the oceans store the CO_2 as a sink, lessening the greenhouse effect and cooling the atmosphere. According to the Intergovernmental Panel on Climate Change (IPCC), during the past 650,000 years, CO_2 levels have closely followed the glacial cycles. Warm interglacial periods had high CO_2 levels, and cold glacial periods had low CO_2 levels.

There is also a human factor that each person needs to consider: Almost everything everyone does contributes to the greenhouse effect (such as driving cars, using electricity, buying any manufactured item, using utilities), and it is up to everyone to work toward a solution to conserve natural resources and use less nonrenewable energy. I want to show you how we all leave a "carbon footprint" for which we need to take environmental responsibility.

In order to go back in time and look at the "ecological footprints" that have been left by specific natural processes, climatologists can reconstruct past climates, study them, and then apply what they have learned to present-day observations. *Global Warming Trends: Ecological Footprints* will focus on how the climate on Earth has fluctuated over time and the various techniques available to scientists today that allow them to study the physical clues left behind by ancient climates and determine when specific changes in climate took place.

Chapter 1 looks at the science of paleoclimatology and how being able to reconstruct the Earth's past climates helps scientists better understand global warming and how it is affecting the Earth today. Chapters 2 and 3 present an overview of significant climatic periods of the Earth's past and introduce the concept of climate proxies—natural indicators that can be used to infer past climate. They also examine the concept of geochronology and how scientists are able to determine the relative ages of objects on Earth. Chapter 4 focuses on the evidence of climate change and global warming that is present in various erosional and depositional features of the Earth's surface and how the study of these landforms and a solid understanding of the processes that created them clues scientists in to what the Earth was like in ancient times. Chapter 5 explores evidence about climate change that can be successfully derived from the ice cores of Antarctica and Greenland and the types of data they contain as well as evidence obtained from marine cores and the preserved secrets held in their sediments at the bottoms of lakes and oceans.

The next chapter dwells on how scientists are able to use pollen, tree rings, plant remains, and other life-forms to make inferences about past climatic conditions. Chapter 7 focuses on how climate has affected the rise and fall of civilizations throughout time, enlightening scientists to the very real issues humans face today with global warming. This volume then looks at two new types of technology and how they are being used to discover the Earth's past climates. It concludes with the most current information from climate experts and how the futuristic role of computer modeling is helping scientists to discover the past and use it to predict the future—something with which you may find yourself directly involved.

The Science of
Paleoclimatology

When climatologists study the current climate, they have a wealth of information at their fingertips. For example, they can obtain data from instruments, such as barometers, anemometers, thermometers, and rain gauges at weather observatories around the world to collect data for rainfall amounts, temperature, evapotranspiration rates, humidity, wind speed and direction, and major flow of air currents, such as the jet stream. Climate data can be collected from the mountains and valleys of all continents, including Antarctica; from the oceans; and from sophisticated satellite equipment in space. Because of this, and especially with the advances in computer technology, science has made great strides in recent years in being able to study, understand, and predict climate. Multiple types of data can be collected. Climatic data for areas can be gathered and put into computer models, such as temperature, precipitation, wind speed and direction, and humidity.

Other types of supporting data—called ancillary data—can also be collected to give scientists a better understanding of the weather. Ancillary data, such as pollution levels, population density, types and amounts of industry, modes of transportation, levels of development, types of dominant farming practices, land use changes, deforestation, and urbanization, are all pieces of information that can be used to provide a clearer understanding of climate change and global warming.

Fairly good records have been kept for the past 150 years by various international and government agencies and academic institutions. Beyond that time frame, however, data become much more scarce, and it becomes much more difficult to study and understand climatic interactions as they have related to change, including global warming.

THE PURPOSE OF PALEOCLIMATOLOGY

Before written records were kept, scientists did not have the convenience or luxury of accessing easily available, ready-to-use data. Instead, they used older, existing data that could have been interpreted in a meaningful way. This is where paleoclimatology comes into play. Paleoclimatology is the study of climate prior to the availability of recorded data, such as temperature data, precipitation data, wind data, storm data, and other measurements of the weather. The word comes from the Greek root *paleo,* which means "ancient," and the term *climate.* Paleoclimate research helps scientists better understand the evolution of the Earth's atmosphere, oceans, biosphere, and cryosphere. It also helps climatologists quantify the various properties of the Earth's climate that force climate change and better understand the sensitivity of the environment to those forcings.

The National Aeronautics and Space Administration (NASA) uses paleoclimate data to test their computer models that attempt to portray climates different from what exists today. By being able to develop computer models that accurately simulate and portray past climatic activity accurately, scientists gain greater confidence in models they are building today to predict future climate scenarios for various places on Earth. If models are accurate on past incidents, then using these same models on current data raises confidence about future predictions. These models are extremely helpful because they are able to model variables that cannot be found in the geologic or fossil records, such as wind patterns, energy transportation, and cloud distribution.

For example, Gavin Schmidt, a climate modeler at NASA's Goddard Institute for Space Studies (GISS), created a climate model based on the atmosphere's response to the 1991 eruption of Mount Pinatubo in the Philippines. The massive amount of sulfur dioxide the volcano spewed into the stratosphere became sulfate aerosols (tiny reflective particles) that encircled the Earth for more than a year after the eruption, shielding it from the Sun's energy and causing the atmospheric temperature to cool by 0.8°F (0.5°C). The initial model worked well with one exception. According to Schmidt, "It turns out that most of the effects were well-modeled—it got cooler by about the right amount, and the water vapor feedback seemed to be well captured. The model, however, had one major flaw. In the winter following the eruption, actual temperatures in Eurasia were higher, not lower, than normal (the rest of the world was cooler). The model failed to reproduce this winter warming. Global climate models, however, do not generally do a good job with the stratosphere—the section of the atmosphere affected by Pinatubo's sulfate aerosols. Because the stratosphere does not influence weather, there are only a few models that have been built to describe it."

Because of this, Schmidt went back and looked closer at a phenomenon called the North Atlantic Oscillation (NAO), which is a permanent pressure system that exists over the Atlantic between the Azores Islands and Greenland. The NAO alternates between positive and negative conditions, and when the NAO is positive, it warms Eurasia, just as Mount Pinatubo's eruption did. Based on this information, Schmidt rebuilt the model, taking the stratosphere's reaction into account, and ran the model again. In order to check his results, he entered data from an era when known stratospheric changes had also taken place: the Maunder Minimum, a period of notable cooling in Europe between 1650 and 1710 when the Sun was relatively quiet. According to Schmidt, "This is an example where paleoclimate and satellite data came together to help scientists build a better model of how the stratosphere influences the NAO. This revised model was able to reproduce the unusually cold temperatures over Europe during the Maunder Minimum and was also able to reproduce the unusually warm temperatures over Europe after the Pinatubo eruption." Schmidt's goal for paleoclimatology is to provide the information needed to validate and refine other models, especially those designed to predict abrupt climate change—a critical issue in light of global warming. According to Schmidt, "We [currently] can't get the models to do some things like rapid climate change. We [humanity] owe our entire history to the fact that that happened, and we don't know why it did" (referring to the abrupt climate change after the Earth's last ice age).

NASA's GISS has been able to simulate various climate sequences throughout the Earth's history, such as the major glacial episodes, especially the Last Glacial Maximum and Holocene, which cover the past 18,000 years. These models can also be used to test for the climate system's sensitivity to change in carbon dioxide levels, a key component of global warming today.

Scientists want to be able to reconstruct past climate to gain a better understanding of what natural variations in climate have occurred over the past several thousand or more years, why they have occurred, and how these variations have affected the environment. It is also helpful in order to gain a better understanding of climate variation independent of human interference (because all historically recorded data have occurred during the time of human disturbance). Paleoclimatology includes both collecting evidence of past climate conditions and striving to understand the processes that caused the conditions—a "cause and effect" relationship.

Another important concept scientists have learned to appreciate from discoveries in paleoclimatology is that the Earth's climate is prone to frequent change. Throughout geologic time, there is evidence of floods, droughts, warm periods, and ice ages. By studying past climatic intervals, scientists are better able to reliably make predictions about how climatic changes will affect the environment and how long and how widespread their effects will be.

One key piece of knowledge of which scientists have gained a better understanding in recent years is that of abrupt climate change. They have been able to detect periods when the Earth was nearly frozen over and other times when it was a literal hothouse. Sometimes climate changes have happened gradually over very long periods of time, and other times significant changes have happened in a matter of decades or even years. It is important to know what kinds of changes are possible in a complex climate system in order to avoid unexpected surprises in light of recent global warming issues. This is one reason why being able to construct accurate models that depict abrupt climate change is so important.

Studying the past "natural" climatic cycles of the Earth also gives scientists a good base level of data against which to compare present-day situations. As humans interact with the atmosphere, they directly affect the climate system. The amount of pollution they add to the atmosphere, for instance, has a direct effect on global warming. By having a firm understanding of the Earth's climate throughout time, scientists can assess the specific effects of natural phenomena on the weather (such as volcanic eruptions) compared to human-induced phenomena (such as pollution, deforestation, and farming practices). One major finding is that the sharp rise in temperature seen in the 1900s is uncharacteristic compared to earlier time periods. Other time periods have not had such a sharp, distinct increase in temperature. A study led by Dr. James Hansen of NASA's GISS along with scientists from other organizations concluded that the Earth is now reaching and passing through the warmest levels it has seen in the past 12,000 years. They concluded that data show the Earth has been warming at the rapid rate of approximately 0.36°F (0.2°C) per decade for the past 30 years.

According to Dr. Hansen, "This evidence implies that we are getting close to dangerous levels of human-made pollution. In recent decades, human-made greenhouse gases have become the largest climate change factor." Even when current temperatures are compared to those of the last documented significant warm period, known as the Medieval Warm Period, which occurred from 800–1300 C.E. in Europe, temperatures today are 0.7°F (0.4°C) higher. This finding tells scientists that the human contribution to present-day global warming is significant and must be addressed if global warming is to be effectively dealt with.

Paleoclimatology assists computer modelers in refining their climate modeling programs. These computer models are extremely complex because the climate system has so many variables involved in it. When programmers design programs using paleoclimatic knowledge, it helps calibrate the models, making them more accurate overall and increasing their predictive power. With the interest today on rising temperatures and greenhouse gas concentrations, understanding the past is a way to compare it to the present and then be able to predict what is to come.

Scientists use several methods to study past climate. The type of method they use depends on how far back in time they want to go. If scientists are only looking backward less than 20 years, they can use available recorded data, including a vast database of satellite data and instrumental weather measurements. (The U.S. National Oceanic and Atmospheric Administration [NOAA] currently maintains this type of data.) As mentioned previously, other recorded data extends back into the 1800s and some other written records go back even further. For example, written records exist from the Middle Ages in Europe that record data on events such as grape harvests for wine making. If it is known what crops were farmed during a certain period, it is possible to make reasonable conclusions as to what the climate was probably like at the time.

This type of data, however, does not give much information on the long-term aspects of climate change. Some changes in climate take place in cycles of thousands or hundreds of thousands of years or even longer. In order to gain a good understanding of the processes that contributed to climate change and the results of it, climatologists must be able to study the records of broad sections of geologic time. Older climate data are also important to obtain because they give climatologists valuable information about natural climatic conditions before the beginning of the Industrial Revolution and subsequent large-scale human interference on climate. The beginning of the Industrial Revolution was about the same time climate records began to be kept. In order to look further into the past, scientists must use what is called proxy data.

Proxy data are simply natural data that can be used as markers, or indicators, about past climate, such as coral, tree rings, and layers of sediment. These items can contain preserved information about the Earth's atmospheric conditions and climate of the past. Proxy data will be dealt with in more detail in chapter 3.

Scientists have been able to learn much about the Earth's past climate through the tools available in climatology. They have been able to successfully determine that the Earth's climate is always fluctuating and has gone through several ice age cycles. Some ice age cycles have lasted thousands of years with glaciers advancing, then retreating. The last major ice age ended about 10,000 years ago. Since then, the Earth has fluctuated but generally warmed, although a Little Ice Age episode extended from approximately 1450 to 1890 C.E. in the Northern Hemisphere. This occurred after a warming period referred to as the Medieval Climate Optimum.

WHAT PREHISTORIC CHANGE REVEALS ABOUT THE FUTURE

An important part of being able to make forecasts of the Earth's future climate is to know how the Earth's climate has varied in the past and what mechanisms have made it vary. If paleoclimatology can help create past climate and give scientists a better idea of how climate has changed throughout time, it gives them important insight as to how today's actions will influence the trend of future climate, especially concerning global warming.

Since scientists have maintained a detailed record of the Earth's climate for the past 150 years, they have been able to determine that the temperature has warmed by 0.9°F (0.5°C). Because of its short duration, however, it is difficult to say how much of this warming can be directly attributed to humans, the burning of fossil fuels, and the enhanced greenhouse effect and how much of it is due to natural variations, such as solar variability and other factors.

Because the issue of global warming remains such a heated debate today among opposing groups, having paleoclimatic data is helpful because it gives scientists a handle on how much of the Earth's climate has naturally varied throughout time (under the influence of volcanic eruptions, orbital changes, and solar output) without any interference from humans.

According to NOAA, paleoclimatology helps scientists find answers to questions such as the following:

- Is the past 100 years of increased temperature and global warming normal?

- If the last century was abnormally warmer than it should have been, what does that mean?
- Is the recent rate of climate change normal?
- Is the recent warming something new or just part of an older, larger cycle?
- Is there any evidence of past climate forcings (outside factors forcing the climate to behave in a certain way) that could be actively contributing to the warming trend today of which we are currently unaware?

By being able to differentiate between natural and human input, approaching the issue of global warming can be much more scientific and yield better solutions. In other words, having a paleoclimatic perspective allows scientists to look back thousands of years to develop a more accurate picture of how the Earth's climate may change and affect everyone in the future.

RATE OF CHANGE

One important concept of paleoclimatology is to be able to track the changes in greenhouse gas concentrations by the heating or cooling of the Earth's surface. Scientists know that when the atmosphere warms up, carbon dioxide is released from the oceans. In addition, if the Earth's orbit changes and triggers a warming period, it can also increase greenhouse gases. When this happens, it triggers the greenhouse effect. This creates a positive feedback that encourages more warming. Conversely, when temperatures cool, CO_2 is absorbed by the oceans, which contributes to additional cooling. According to the Intergovernmental Panel on Climate Change (IPCC) in their 2007 report, during the past 650,000 years, CO_2 levels have been high, and during the cool glacial periods, CO_2 levels have been low.

Because scientists can observe coinciding rates of change with temperatures and CO_2 levels throughout the past several hundred thousand years and determine that there is a strong correlation, this also gives them a scientific basis and better understanding about the future. Through the correlation of temperatures and CO_2 records, scientists can determine, through modeling, the types of effects that can be expected from various levels of CO_2 in the atmosphere.

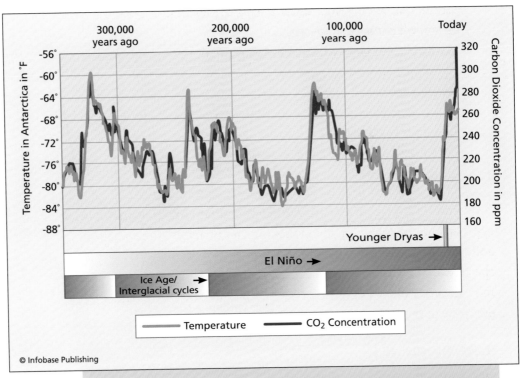

Fluctuations in temperature (red line) and in the amount of carbon dioxide concentrations in the atmosphere (blue line) over the past 350,000 years. The temperature and carbon dioxide concentrations at the South Pole run roughly parallel to each other, showing the strong correlation between the two.

Ocean currents are also affected by changes in the Earth's surface temperature. It has been proven that melting glaciers and ice caps can add enough freshwater to the northern oceans to slow or stop the flow of key currents such as the Gulf Stream. Because ocean currents have a significant effect on climates around the world, these changes can cause significant changes to world climate. For instance, if the Gulf Stream were to slow down or stop due to excess melting of glacial and ice cap ice in the Arctic, it could bring cold, ice age–like weather to normally moderate temperature regions in Europe. If ocean currents are altered, the global distribution of heat will be altered, which will cause major changes in climate from one area of the Earth to another.

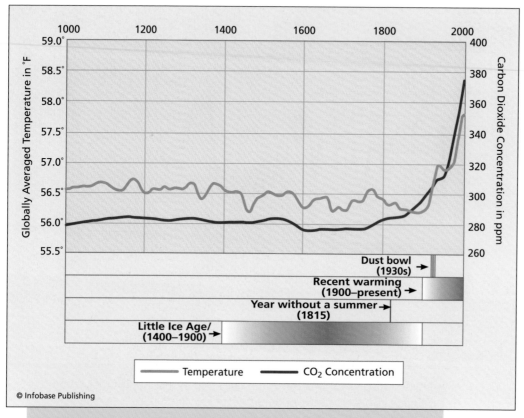

This graph shows how the recent warm temperatures and rising carbon dioxide levels relate to those of the past 1,000 years. Experts believe this sharp rise is due largely to how humans have enhanced the greenhouse effect.

Scientists have gathered information from records preserved in layers of ocean sediments of the fossils of marine life that show that this has happened several times throughout the Earth's history and that these changes have caused significant climate changes over large areas.

NOAA has completed several studies that show that the Earth's earlier climate had periods of stability interrupted by periods of rapid change. They believe this is because interglacial climates (such as the current climate) are more stable than cooler, glacial climates. Disruption of this stability can be caused by variations in the Earth's orbit through time and the associated variability of solar radiation received at

This rock landscape of the American Southwest was shaped over eons of geologic time by the multiple forces of wind, water, and ice. *(Nature's Images)*

the Earth's surface. Glacial periods occur when summer solar radiation is reduced in the Northern Hemisphere. Throughout the Pleistocene epoch (the past 1.8 million years), these cycles have occurred with a frequency of about 100,000 years. Precession of the equinoxes also affects solar radiation in cycles of 23,000 years, which encourages glacial periods. Warming at the end of glacial periods also happens abruptly due to the ice-albedo feedback mechanism. With less ice to reflect incoming solar radiation, more heat is absorbed by the Earth's surface, causing temperatures to rise. Once ice begins to melt and expose the land and water, additional solar radiation can be absorbed by the Earth's surface, raising temperatures and causing even more ice to melt in a positive

Clues to past climate can be seen in present-day landscapes. The basin this alpine lake is contained within was originally carved out by a glacier during the last ice age. The lake is in the Uinta Mountains in northern Utah, part of the Rocky Mountains. *(Nature's Images)*

feedback. Abrupt, rapid climate changes often accompany the transitions between glacial and interglacial periods. Most current civilizations came into existence during relatively stable periods of climate.

CLIMATE PATTERNS

Climatologists have determined that climate change is not only constant but that it also occurs at multiple time scales. It occurs at geologic time scales of millions of years, and nested within that at scales of hundreds of thousands of years, and within that at tens of thousands of years to thousands, hundreds, and even decades and annual scales. Long-term changes are seen in changes in the Earth's tilt, axis, and precession. Cycles are seen in ice ages. Shorter-term changes are seen in patterns of phenomena, such as El Niño. Historical climate change will be looked at in more detail in chapter 2.

Some scientists have even suggested that climate has a "memory" and that by understanding past conditions and behaviors it is possible to

Bryce National Park, Utah. This area was once covered by seas, mountains, deserts, and coastal plains. The geologic formations within this area testify to the various climates that existed when each layer was deposited. *(Nature's Images)*

predict and plan for future conditions. According to scientists at NOAA, "Understanding 'climate surprises' of the past is critical if we are to avoid being surprised by abrupt climatic change now." For example, abrupt climate change has had drastic consequences for ancient civilizations that can serve as warnings for society today. According to Dr. Harvey Weiss, a professor of archaeology at Yale University, as reported by CNN in April 2005, climate change was a fact of life for earlier civilizations. Egyptian pharaohs and medieval Vikings both had to deal with violent changes in weather patterns, which sometimes prompted mass migrations.

According to Dr. Weiss, "Those episodes proved to be the single most important stimulus for the major transformations in human history." As information is discovered about prehistoric climates, archaeologists are looking for connections between climate change and human development. They have linked the collapse of early Bronze Age civilizations in Greece and India to abrupt climate changes about 4,200 years ago. Drought has been cited as a factor in the collapse of the Anasazi civilization of the American Southwest during the 13th century.

Dr. Weiss says results were not always negative, however—it was drought that drove farmers in ancient Mesopotamia to build irrigation channels, which enabled farmers to grow enough food so that for the first time everyone did not need to farm. Instead, it allowed some inhabitants to pursue other paths, such as architects, politicians, and artists. According to Weiss, "The historical lesson is that those societies had no knowledge of what was happening to them and certainly no historic knowledge of what could happen to them, where we have both." With today's technology, scientists are able to improve predictions for the future based on knowledge learned from incidents in the past.

Because different aspects of the environment respond in different ways, physical characteristics of areas play a significant role in determining climate. As an example, if a highly vegetated area were to have repeated years of lower-than-normal rainfall, the soil would dry out, lakes would shrink, and water sources would diminish. In response, the vegetation would die off. Over time, with less vegetation, less biomass would enter the soil, causing the soil to lose fertility. There would be less evapotranspiration, which would also change the characteristics of the area to become more arid. Evapotranspiration is the transfer of moisture from the Earth to the atmosphere through evaporation of water and transpiration from plants. Dry conditions would contribute to even further evaporation and shrinkage of lakes, cause grasslands to dry up, and cause desert areas to begin expanding. This is a process called "desertification" that is becoming common in the American Southwest as global warming continues, and it has caused enormous problems for the peoples of Africa. It has destroyed water supplies, drinking water, and farming and has subsequently contributed to the spread of disease and starvation. Desertlike conditions can then lead to less rainfall in subsequent years. When this works as a cycle, scientists view this as a self-perpetuating process. This happens because different parts of the environment respond over different time intervals and in different ways, sometimes feeding off of each other and either enhancing or minimizing the responses.

Some scientists in England have noticed that there is a correlation of summer rainfall amounts to the wintertime air pressure pattern of the North Atlantic Ocean; the summer climatic pattern seems to "remember" what the previous winter pattern was. Because of this correlation, farmers have been able to predict which years will be better for growing wheat, because they have been able to predict which years will have wetter or drier summers.

Being able to look at characteristics of climate in certain locations and to predict how climate will act in the future is a valuable tool. One thing that climatologists agree on is that the Earth's climate is always changing. Its variability changes on multiple timescales—short-, medium-, and long-term intervals. No matter what the scale of these changes, they have a significant effect on humans. Because of this, it is important that scientists study past climatic variability in order to better understand future climate change and the potential impacts it may have on society.

Key Climate Intervals in the Earth's Past

This chapter looks at the Earth's climate over geologic time and demonstrates when key long- and short-term changes took place. Climatologists have been able to reconstruct a representative climatic time line by tapping into natural storehouses of data, such as corals, ocean and lake sediments, ice cores, fossil pollen, tree rings, and other physical clues left behind from natural processes. These types of data, known as proxy data, are natural records of climate variability. Through the analysis of proxy records, scientists can extend their understanding of past climate backward in geologic time. This chapter first looks at the Earth's most ancient climates (2 million to 4.6 billion years ago), then progresses to the Pleistocene climates (10,000 years to 2 million years ago), and finally looks at the Holocene/recent climates. It also addresses what climatologists have identified as the key uncertainties.

THE EARTH'S GEOLOGIC PAST

Being able to look to the past has helped climatologists better understand long-term processes relating to the Earth's climate. Although lack of usable data means that much of the most ancient time intervals is not understood as well as are more recent time intervals, as new discoveries are made and analysis techniques are improved, the Earth's past climate history becomes clearer.

In order to put the Earth's past in a historical context, it must be looked at on a timescale that spans 4.6 billion years—what geologists refer to as the geologic timescale. The Earth's history has been divided into major time intervals (see page 166 for Geological Timescale), the intervals determined by significant events in the history of the Earth, such as extinctions, glaciations, and changes in dominant life-forms; which is why the intervals all represent different lengths of time instead of preset consistent intervals. For example, the division between the Permian and Triassic periods (also the boundary between the Paleozoic and Mesozoic eras) corresponds to the Earth's largest mass extinction. At this point, 100 percent of the trilobites (a marine arthropod) were rendered extinct, 50 percent of animal families were lost, 95 percent of all marine species died, and many species of trees were lost. Geologists have determined this could have been caused by glaciation or volcanism.

The geological timescale consists of four division types: eons, eras, periods, and epochs. Eons are the largest intervals of geologic time—they can be hundreds of millions of years long. Eons are subdivided into eras. Eras are then subdivided into periods or even smaller epochs. Epochs are used for relatively young geologic deposits. This hierarchical classification scheme divides time periods based on significant events.

When looking at paleoclimate change throughout geologic time, scientists look at factors such as changes in solar output, changes in the Earth's orbit, precession, and tilt, changes in the positions of the Earth's continents, and changes in the atmospheric concentration of greenhouse gases. The following table illustrates what general temperature regimes and life-forms existed during different geologic periods.

Characteristics of Geologic Periods			
GEOLOGIC PERIOD	AGE (MYA)*	GENERAL TEMPERATURE	DOMINANT LIFE-FORMS
Cambrian	550–480	Mild	Trilobites
Ordovician	480–415	Cold through hot	Plants, corals, fish
Silurian	415–390	Warm	Insects, vascular plants
Devonian	390–345	Hot	Land plants, amphibians
Mississippian	345–300	Warm	Winged insects
Pennsylvanian	300–270	Mild	Reptiles, ferns
Permian	270–230	Cooler	Amphibians, trees
Triassic	230–200	Hot	Dinosaurs, mammals
Jurassic	200–135	Hot	Dinosaurs, birds, conifers
Cretaceous	135–60	Warm	Dinosaurs, snakes, butterflies, marsupials
Tertiary	60–2.6	Cool	Mammals, grasses, birds
Quaternary	2.6–present	Warm	Humans

million years ago

Concerning long-term change, scientists at the United States Geological Survey (USGS) believe it has been controlled primarily by plate tectonics and its influence on the atmospheric greenhouse effect. In plate tectonics, the Earth's continents "drift" on top of a fluidlike layer of the crust over geologic time. When the plates collide, some of the plate material is subducted (pushed under) the Earth's crust. The subducted crust

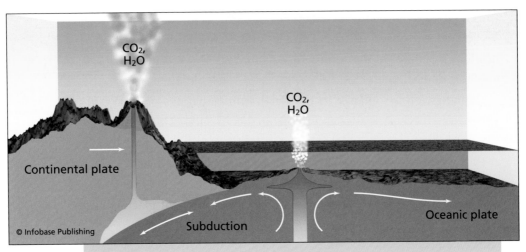

CO$_2$,
H$_2$O

CO$_2$,
H$_2$O

Continental plate

Oceanic plate

© Infobase Publishing   Subduction

Plate tectonics provides one mechanism for carbon dioxide to be
released to the hydrosphere, biosphere, and atmosphere.

can melt, forming a volcano on the Earth's crust directly above. Plate tectonics is responsible for releasing CO$_2$ into the atmosphere. Having an
understanding of plate tectonics has allowed climatologists to study and
date the rocks and landforms on the Earth's surface and recreate models
of where the continents were at different periods in time.

Some climatologists have suggested ancient periods of glaciation
may have even been correlated with plate tectonics. The idea has been
discussed that there may have been a drop in the production rate of
CO$_2$ by volcanoes, which would have lowered the CO$_2$ content in the
atmosphere, thereby lowering the Earth's temperature and making it
prone to glaciation. Some scientists have suggested that global warming may result from rapid plate spreading and that global cooling may
result from slowed plate spreading during the tectonic process.

There are also mid-term changes. These changes focus on what scientists refer to as the "astronomical theory of climate change" and concern the Milankovitch cycles. Named after Serbian astronomer Milutin
M. Milankovitch, the Milankovitch cycles explain the changes in the
Earth's seasons that result from changes in the Earth's orbit around the
Sun. There are three types of orbital changes: (1) changes in the Earth's
tilt, (2) changes in its eccentricity, and (3) changes in its precession. As

the Earth revolves in its orbit around the Sun, these three separate cyclic movements—whose lengths of cycle all differ in duration—combine to produce variations in the amount of solar energy that reaches the Earth and help determine its climate.

The first cycle—change in the Earth's tilt—operates on a 41,000-year cycle. The Earth's tilt can vary from 22° to 24.5°. The smaller the tilt, the less seasonal variation there is between summer and winter at the middle and high (polar) latitudes. Today, the Earth's tilt is 23.5 degrees. When the tilt is smaller, winters are milder and summers are cooler, creating optimal conditions for the formation of glaciers and ice sheets.

Once glaciers and ice sheets build up, then positive feedbacks in the climate system come into play. A positive feedback is an interaction that amplifies the response of the system to what it is subjected to—it increases the condition (such as makes it colder, makes it hotter, etc). In this case, when the Earth is covered with more snow and ice, it reflects more of the Sun's energy into space, which causes additional cooling to occur. Scientists have also determined that the amount of CO_2 in the atmosphere falls as ice sheets grow, which cools the climate further (another positive feedback).

The second cycle—eccentricity—deals with the shape of the Earth's orbital path around the Sun. The orbital path is not a perfect circle. It shifts through time and becomes more or less oval. This means that the Earth is slightly closer to the Sun at some times of the year than others. In a cycle of 100,000 years, the orbital path varies from nearly circular to very elliptical (oval). When the orbit is more circular (as it is today), the seasonal change in solar energy is not great (approximately 7 percent), but when the orbit is highly eccentric, the seasons are much more exaggerated (approximately 20 percent) and the lengths of the seasons are changed. This cycle affects the relative severity of summer and winter and helps control the growth and retreat of ice sheets.

Cool summers in the Northern Hemisphere, where most of the Earth's landmasses are located, allow snow and ice to exist until the next winter, which enables large ice sheets to develop over hundreds to thousands of years. Warm summers shrink the ice sheets by melting more ice than can accumulate during the winter.

The third cycle—change in precession—deals with the spin of the Earth on its axis. Like a spinning top that begins to wind down and wobble as it traces a small circular path, the Earth's axis does the same thing. This wobble is called precession, and it operates on a cycle of around 23,000 years. Because of this, the northern spin axis traces a circle through the nighttime sky. Currently, the Earth's North Pole points at Polaris, the North Star, but as the spin axis wobbles, it will not always point toward Polaris. At this time, the Earth is closest to the Sun in January and farthest away in July, but because of precession, in approximately 11,000 years, the opposite condition will exist, which will give the Northern Hemisphere more severe winters.

Paleoclimatologists can use this information to look at the climatic record over the past million years to identify any cause-and-effect relationships. They have been able to determine that there is a correlation between periods of low eccentricity (circular orbit) and glacial periods. In addition, the interglacial periods (the intervals between glacial periods) over the last 160,000 years show the cyclical patterns of the 41,000-year tilt cycle and the 23,000-year precession cycle.

According to Dr. Perry Samson, a meteorologist at the University of Michigan's Department of Atmospheric, Oceanic, and Space Sciences, other paleoclimate factors exist that also correlate with the Milankovitch cycles and assist in the reconstruction and understanding of past climates. They include factors such as:

- the amount of dust in the atmosphere
- the reflectivity of the ice sheets
- the concentration of greenhouse gases
- characteristics of clouds, and
- the rebounding (uplifting) of land that has been previously depressed by the enormous weight of glaciers and ice caps.

Once the weight is removed, the ground slowly rises to the level it was before the weight of the ice pressed on it—a phenomenon called isostatic rebound.

A good understanding of the Milankovitch cycles has greatly helped scientists put together pieces of the puzzle to explain the advance and retreat of ice over the last 10,000 to 100,000 years.

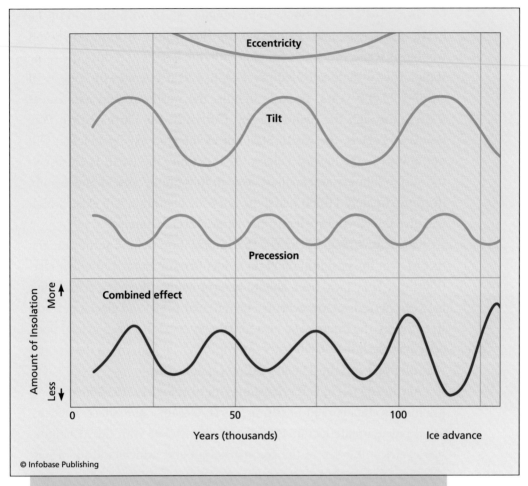

The Milankovitch cycles are one method used to correlate cycles of the Earth's warming and cooling periods. The orbital patterns combine to vary the amount of incoming solar radiation (insolation) reaching the Earth at a given time. One theory is that ice ages are caused by changes in insolation.

ANCIENT CLIMATES—2 MILLION TO 4.6 BILLION YEARS AGO

Climate change of ancient Earth has been determined primarily through the study of plate tectonics and the reconstruction of the locations of the Earth's landmasses, oceans, and waterways as well as geologic evidence of atmospheric CO_2. Much of the paleoclimatic reconstruction of the

Earth's ancient climate has been accomplished through the mapping of past positions of the continents and plotting of the distributions of rock types that form in specific climate regions. Certain formations, such as coal, need certain conditions in which to form. Coal needs abundant rainfall and generally forms in tropical rain forest or temperate forest conditions, which provide the necessary biomass, heat, and moisture.

In general, sedimentary rocks form in areas where water is present. Through a knowledge of rock types and the physical conditions necessary for formation, it is possible to step back in time and reconstruct the physical conditions that must have existed on the Earth at the time the rock was formed. By mapping the past distribution of thousands of rock types, scientists have been able to map the distribution of ancient climatic belts. Over the last 2 billion years, the Earth's climate has been constantly changing between hot and cold conditions, as shown in the figure.

The Earth's most ancient climates include the following geologic time span:

Precambrian eon	570 MYA–4.6 billion years ago
Phanerozoic eon	0–570 MYA
Palaeozoic era	240–570 MYA
Mesozoic era	65–240 MYA
Cenozoic era	2–65 MYA

Within this geologic time span, the Earth experienced a wide range of climate variability. At a period around 635 million years ago, many scientists believe the Earth was completely frozen over into what climatologists refer to as Snowball Earth. After that, 300 million years ago the planet became Hothouse Earth (also referred to as the mid-Cretaceous Greenhouse World).

The Precambrian—the oldest period on Earth—covers about 85 percent of the Earth's history, but because of the extreme time span, there is not much reliable evidence. Scientists have found evidence of the following two major periods of glaciation: one at 2.3 to 2.7 billion years ago and another at 0.9 to 0.6 billion years. The three intervals during this time period that scientists do have information on and that have provided some insight to the Earth's past climate are next

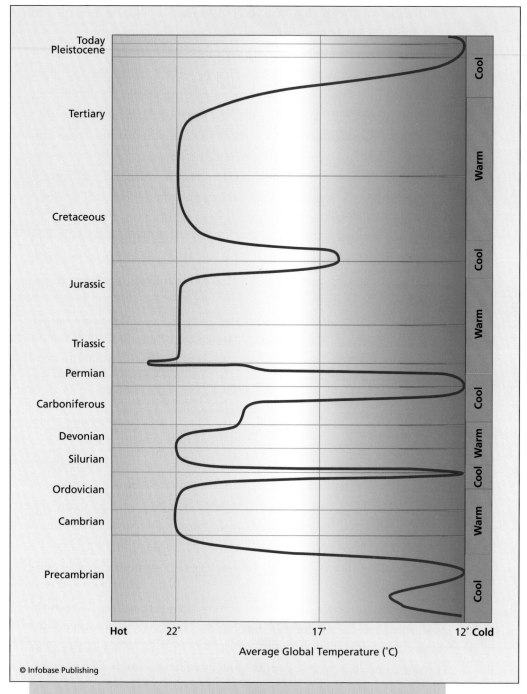

The graph illustrates the way climate has varied over geologic time from hot to cold; today, the Earth is in an interglacial period.

discussed: (1) the Faint Young Sun Paradox, (2) Snowball Earth, and (3) Hothouse Earth.

The Faint Young Sun Paradox

The inability to find more evidence for early glaciation has always stumped scientists because they expected to find evidence supporting abundant glacial events. By studying the evolution of stars in the universe, astronomers have been able to recreate the history of the Sun. The models they have developed indicate that the early Sun was about 30 percent less bright than it is today. These calculations presented a mystery for climatologists, because a decrease of just a few percentage points in the Sun's present strength would cause all the water on Earth to freeze, even with all the CO_2 in the air and the greenhouse effect. If all the water in the oceans, lakes, and streams froze today, their high albedo (reflectivity) would make it difficult to melt the ice. What presents such a mystery to scientists is that with such a weak Sun, even if the Earth's greenhouse gas levels were then what they are today, the Earth should have remained completely frozen for the first 3 billion years of its existence. Yet scientists have not been able to find any evidence to support the Earth ever having been completely frozen. Geological evidence for this time has been partly found in sedimentary rocks, and sedimentary rocks are formed from running water, not frozen water. Evidence of a continued presence of life on Earth during this time also does not support the possibility of a planet frozen because of too weak a Sun. This question remains a mystery.

Snowball Earth

There is one theory concerning the Earth being in a nearly frozen state 635 million years ago that climate scientists debate—an episode referred to as Snowball Earth. The term *Snowball Earth* describes the coldest state in which a planet can exist. In order for this to happen, the global mean temperature would have to be -74°F (-50°C). Most of the solar radiation would be reflected back into space by the high albedo of the snow and ice covering the planet.

There is evidence that supports this hypothesis during this later time period. Evidence exists in sedimentary rocks containing mixtures

Climate scientists believe that the Earth may have been frozen 635 million years ago. *(NOAA, Ardo X. Meyer, photographer)*

of coarse, unsorted boulders and cobbles mixed with fine silts and clays. These unsorted deposits are characteristic of ice and glacial deposition, and they are found on almost all the continents on Earth. Based on evidence found in sedimentary rocks, it has also been proposed that between 550 and 850 million years ago, two to four of these separate ice house incidents may have occurred.

The biggest debate by scientists on this subject concerns the geographic location of the continents. If some of the continents were located in the Earth's equatorial region (the Tropics), this supports the hypothesis that the Earth could have been entirely frozen. If, however, all the continents were located at high (polar) latitudes, they could all have been frozen, but the Tropics could still have remained unfrozen, meaning the entire Earth did not necessarily freeze, although large portions could have. Scientists have done much work to determine the geographic location of the continents.

Research conducted by Paul F. Hoffman (a field geologist) and Daniel P. Schrag (a geochemical oceanographer) of Harvard University has helped to answer many of the questions surrounding this notable climate event. One of the initial enigmas was the occurrence of glacial debris found near sea level in the Tropics. This evidence contradicted evidence seen today—glaciers near the equator now survive only at 16,404 feet (5,000 m) above sea level or higher. Even at the coldest segments of the last great ice age, glaciers did not form lower than 13,123 feet (4,000 m) in elevation. What Hoffman and Schrag found, however, was not only glacial debris near sea level but that it was mixed with unusual deposits of iron-rich rock. What made this odd was that those rocks should have been able to form only in an environment that had little or no oxygen in its atmosphere or oceans. According to scientific evidence, however, the Earth's atmosphere at that time should have closely resembled that of today. Even more puzzling was that there were also deposits of rocks that could have formed only in warm water found in the rock layers that formed just after the glaciers receded. This presented a puzzle: If the Earth were cold enough to ice over completely, how did it warm up again, especially to such extremely hot conditions? In addition to that conundrum, the carbon isotopic signature in the rocks hinted at a prolonged drop in biological production, leaving scientists to conclude that there had been a dramatic loss of life at that time in the Earth's history. Hoffman and Schrag make sense of these enigmas, however, in their field studies, as reported in an article they presented in the journal *Science* in January 2000.

Based on his discoveries and work in 1964 concerning the magnetic orientations of mineral grains in glacial rocks, W. Brian Harland of the University of Cambridge believed that the Earth's continents had all clustered together near the equator. Because he realized that glaciers must have covered the Tropics, Harland was the first geologist to propose the concept that the entire Earth had experienced a great ice age event. While Harland was busy with his research and trying to figure out just how glaciers could have survived the tropical heat, physicists were beginning to develop the first basic mathematical models of the Earth's climate system. In particular, Mikhail Budyko of the Leningrad

Geophysical Observatory discovered a way to explain this enigma. He developed a series of equations that described the way solar radiation interacts with the Earth's surface and atmosphere to control climate. As snow and ice accumulate on the Earth's surface, their high albedo cools the atmosphere and stabilizes and perpetuates their existence. What Budyko referred to as ice-albedo feedback is the same mechanism that helps modern polar ice sheets grow.

An interesting thing occurred with his experiments, however. His climate simulations found that the ice-albedo feedback can get out of control, which is what happened to cause Snowball Earth. When ice formed at latitudes lower than 30° north and south of the equator, the Earth's albedo began to rise at a faster rate because direct sunlight was striking a larger surface area of ice per degree of latitude. This caused the feedback to become so strong in his simulation that surface temperatures dropped severely, which quickly caused the entire Earth to freeze over.

At first Budyko was puzzled at his results, reasoning that if the entire Earth had frozen over, then it must have killed all life on Earth. Yet when scientists examined rocks that were 1 billion years old, they found microscopic algae that resembled modern forms, leading them to believe that life did not cease during this time. Budyko, along with other scientists, was also hesitant to take his model too seriously because he also thought that if the Earth had entered a runaway freeze, it would not be able to pull itself out of it.

The scientific attitude toward these questions began to change in the 1970s, however, when communities of organisms living in places once thought too harsh to allow life to survive were discovered. Seafloor hot springs today support microbes that thrive on chemicals instead of sunlight. This clued in Budyko and other climate scientists to the fact that during Snowball Earth, the volcanic activity that feeds hot springs would have continued to function and could readily have supported life.

The explanation for why the runaway freeze stopped was also answered with newly discovered evidence—it all hinges around CO_2. In 1992, Joseph L. Kirschvink, a geobiologist at the California Institute of Technology, determined that during Snowball Earth, the planet's shifting tectonic plates continued to produce volcanoes above subduction zones, releasing CO_2 to the atmosphere. While the CO_2 was collecting in the atmosphere, there was no rainfall to erode rocks and bury carbon (because the water was frozen), thereby allowing CO_2 levels to become extremely high. CO_2 also entered the oceans through subsea volcanoes and vents. Slowly over the years, atmospheric CO_2 built up and increased the radiative forcing due to the greenhouse effect. Eventually, temperatures at the equator reached the melting point, and the dark surface melt waters caused more solar radiation to be absorbed, soon melting and exposing even larger areas of meltwater. These feedbacks started the process of melting back the ice holding the planet in its grip and conceivably took only a few thousand years to recover from Snowball Earth. Therefore, Snowball Earth was ended by a large-scale intensified greenhouse effect—a large-scale global warming event.

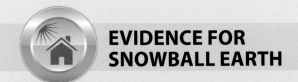

EVIDENCE FOR SNOWBALL EARTH

Scientists have found several pieces of evidence that support the existence of Snowball Earth, such as the following.

- global distributions of glacial deposits on all continents
- land areas that would have been near the Earth's equator at the time have glacial deposits on them
- evidence of flooding and water flow exists where land would have been pushed down under the weight of glaciers then uplifted when the heavy ice melted and water flowed off of it
- glacial marine deposits occur in areas where the warmest surface parts of the ocean are

Evidence of Snowball Earth also brings up the issue of rapid climate change and its importance to life on Earth today. As evidenced from this event, CO_2 plays a critical role in the Earth's climate. As humans continue to affect the climate by heating the atmosphere with greenhouse gases, rapid climate change is a serious possibility—one that could have far-reaching ill effects on humanity and the environment.

The Phanerozoic is composed of three eras—the Palaeozoic, Meso-zoic, and Cenozoic. It was during this time that the landmasses of the Earth came together in the supercontinent called Pangaea. Scientists believe that during the Palaeozoic there was increased volcanic activity and that atmospheric CO_2 in the early part of this period may have been high based on evidence geologists have found in carbonate minerals from the time. Some estimates have been made that the atmospheric concentration of CO_2 may have been 10 times greater in some locations than it is today.

Hothouse Earth

During the mid-Cretaceous period, 90 to 120 million years ago, the Earth was much warmer than today. Often referred to as Greenhouse World, the Earth's warmth extended even to the high (polar) latitudes. This evidence is supported by the abundance of fossil records of plants and animals at polar locations that are found only in warm environments. For instance, corals were discovered far from the equator. Warm-water animals and plants have also been found in polar locations. Scientists have determined that breadfruit trees, a species of tropical vegetation native to the Malay Peninsula and western Pacific islands, grew even in Greenland.

In scientists' efforts to reconstruct past continental positions and ocean currents, they have determined that the Earth's geography and ocean currents were different during the Hothouse Earth period. They have also determined that CO_2 levels were much higher—up to four times higher than today. This Hothouse Earth followed the Snowball Earth conditions mentioned previously, so the shift was due to both the rising atmospheric CO_2 concentration and the melting of the vast ice sheets.

According to geologists at the University of North Carolina, this period of intense warming triggered the greatest mass extinction in the Earth's history about 250 million years ago, as opposed to the theory that the mass extinction was caused when a meteor hit the Earth. Instead, they claim that volcanic eruptions and global warming were the causes of the mass extinctions during the Permian. As a result of their research, they have determined that the climate turned exceedingly hot because

Following Snowball Earth, the Earth warmed up dramatically and became much warmer than it is today during the Mid-Cretaceous period 90 to 120 million years ago. Vegetation even grew in the polar regions. *(Nature's Images)*

of a large increase in volcanic activity that released huge amounts of CO_2 and methane into the atmosphere, thereby causing rapid global warming. Also harmful to life on the planet was the enormous amount of hydrogen sulphide that entered the atmosphere, which damaged the ozone layer and killed the majority of life-forms.

According to Christopher Poulsen, a paleoclimatologist at the University of Michigan, as the ice melted from the ice-covered areas and the atmospheric CO_2 concentration rose, the Tropics became much more arid, the vegetation became stressed, and the ecosystem was replaced by desert. One of the most important things he learned from his research is that the Tropics are very susceptible to large climate changes.

Another episode occurred 55 million years ago when methane was released from wetlands and turned the Earth into another hothouse. Methane is a very powerful greenhouse gas; it is 23 times more effective than CO_2. This event is referred to as the Paleocene-Eocene Thermal Maximum (PETM). During this time, the Earth's surface warmed 9°F (5.4°C) in a few hundred years. The Earth warmed up so much that the Arctic Ocean even reached a temperature of 73°F (23°C). Sea surface temperatures rose 8.3–13°F (5–8°C). Scientists support the theory that the entire ocean depths heated up, not just the surface. In addition, the chemical composition had also changed and become harmful. Oxygen content was drastically reduced, causing major die-offs of deep-sea foraminifera.

Climatologists consider this period one of the most significant examples of Earth's sudden global climate change. The period is an example of what an extreme, rapid global warming event causes. It also coincides with a major extinction of both ocean and land species. This major global warming period lasted for about 200,000 years. Life on land was replaced with animals that could survive the extreme temperatures. These animals were mainly smaller versions of the mammal groups that exist today.

THE PLEISTOCENE CLIMATES—10,000 YEARS TO 2 MILLION YEARS AGO

There is much more geologic evidence for the younger geologic ages, such as the Quaternary period, which spans the last 2 million years of the Earth's history. During the Quaternary, global climate has alternated between times of warmth and frigidity—interglacial and glacial episodes. The Quaternary is separated into two epochs, the Pleistocene (10,000 years to 2 million years ago) and the Holocene (10,000 years to the present).

The Pleistocene is best known for its glacial and interglacial intervals. Geologists have been able to find an abundance of evidence with which to reconstruct the geologic past during this time frame. Through the analysis of deep-sea sediment cores and windblown loess deposits on land, geologists have been able to determine when, where, and how long glaciers advanced across the Earth's surface. Loess is silt that has been deposited by wind action. It usually forms near the borders of continental glaciers and is the material eroded by glacial meltwater and then transported by wind.

One of the more helpful techniques used to study past climate change is analysis of oxygen isotopes. By studying fluctuations in oxygen-18 (^{18}O), it is possible to calculate past ocean temperature and global ice volume. The ^{18}O ratios are studied in ice cores obtained by scientists in places such as Greenland and Antarctica. The ^{18}O trapped in the ice gives climatologists a clear indication of what climate was like when it was deposited. Many scientists believe the glacial cycles during this period were driven by influences from the Milankovitch cycles—changes in the Earth's tilt, eccentricity, and precession.

Variations in atmospheric CO_2 are considered another major factor in determining the climate during the Pleistocene. Climatologists agree that a strong correlation exists between temperature and CO_2 levels, that is, that the greenhouse effect has played a significant role in the Earth's global climate, especially over the past 160,000 years. A redistribution of carbon took place among the various carbon reservoirs on Earth—the atmosphere, ocean, land, and biosphere—but the exact relationship that triggered the glacial and interglacial intervals is not clearly understood. Three notable events occurred during this time period: (1) the Penultimate Interglacial period, (2) the Dansgaard-Oeschger events, and (3) the Heinrich events.

Penultimate Interglacial Period

The Penultimate Interglacial period occurred about 125,000 years ago, and at that time Northern Hemisphere summers were slightly warmer than today, roughly 1.7–3.3°F (1–2°C) warmer. This interglacial is also referred to as the Eemian interglacial and is attributed to changes in the Earth's orbit.

One interesting thing to note is that because climatologists consider this period to have had similar temperatures to those we are currently facing with global warming, they have used the geologic record to reconstruct environmental conditions in an attempt to understand the types of conditions humans will be facing in the coming years with global warming. What they were able to reconstruct is that at that time, with the increase in temperature of 1.7–3.3°F (1–2°C), sea levels rose about 13–19 feet (4–6 m) above the levels of today, and the majority of southern Greenland's glaciers had fully melted. Because these glaciers melted, they contributed 6.6–9.8 feet (2–3 m) of the rise in sea level, which is a vivid example of why the glaciers that are melting today in Greenland, the Arctic, Antarctica, and other continents are of such concern to climatologists.

Scientists have run paleoclimatic models on this data and have been able to determine that this major melting of glaciers was triggered by temperatures that were only 5–8.3°F (3–5°C) warmer than today. What is also alarming to climatologists is that the Penultimate Interglacial period's physical conditions are the closest to the physical conditions of today.

Dansgaard-Oeschger Events

A notable series of rapid fluctuations of temperature that occurred during the last ice age is called the Dansgaard-Oeschger events. Through the analysis of ice cores and the reconstruction of temperatures, it has been determined that there were 23 fluctuations between 110,000 and 23,000 years ago. The cyclic warming periods in the Northern Hemisphere all show a distinctive pattern: The atmosphere warmed rapidly (generally in a few decades) then gradually cooled over a longer period of time. Although a multitude of ice cores have been studied for this time period, scientists still are not sure what caused the cyclic behavior of the atmosphere. The behavior of the cycles was different in the Southern Hemisphere, however. The warming cycles tended to be much more gradual with less temperature variation.

Heinrich Events

Some scientists have suggested that the Dansgaard-Oeschger events are related to the Heinrich events. Hartmut Heinrich, a marine geologist, described these events as episodes during the last ice age when major icebergs cleaved off ice sheets and drifted to sea. His proof of this was found in ocean-bottom sediment cores that contain the eroded sediments the ice had originally scoured and carved off the landmasses as it moved over land. He claimed that as the icebergs melted and added freshwater to the oceans, they interrupted the major ocean currents. In particular, he believed they interrupted the Atlantic thermohaline current, the key current that transports heat from the equatorial area to the

North Atlantic. The addition of freshwater slows the current, destroying its heat-carrying capacity and significantly cooling the Northern Hemisphere. This, in turn, was enough of a change to cause the cycle of warming and cooling seen in the Dansgaard-Oeschger events because cooler climate causes the ice cover to increase. When ice cover increases, so does surface albedo, which, as a positive feedback, promotes further ice growth and additional cooling.

THE HOLOCENE/RECENT CLIMATES—10,000 YEARS AGO TO PRESENT

The Holocene brings geologic time up to the present day. The major ice sheets of the last ice age reached their maximum extent about 18,000 years ago and then began retreating around 14,000 years ago. However, this time span was not one of a continual upward trend in temperatures. The climate varied from significant cooling to thermal episodes. There were four notable periods of climatic cooling or warming during this period: (1) the Younger Dryas Cooling, (2) the Mid-Holocene Thermal Maximum, (3) the Late Holocene Neoglacial Fluctuations, and (4) the Little Ice Age.

Younger Dryas Cooling

One of the more notable periods of cooling is known as the Younger Dryas Cooling. This period occurred 10,000 to 11,000 years ago, which coincides with the geologic boundary of the Pleistocene and Holocene epochs. Glacial ice advanced from the north and moved southward to about 45°N latitude (roughly the same latitude as Minneapolis, Minnesota, and Portland, Oregon). This was a significant advance, because ice levels proceeded within about 10° of where they had reached during the last glacial maximum.

Mid-Holocene Thermal Maximum

The Mid-Holocene Thermal Maximum occurred about 6,000 years ago. Paleoclimatologists have determined that this period was somewhat warmer than it is today. Scientists have uncovered a significant amount of evidence relating to this period and have made great strides in understanding the global issues concerning the cause and extent of temperature change. Changes in the Earth's orbit eventually combined to alter the amount of solar radiation that was reaching various areas of the Earth, according to NOAA. They have concluded that temperatures above summer temperatures today in the Northern Hemisphere did occur. Other evidence that supports the occurrence of this event includes vegetation zones that migrated to different latitudes and changes in precipitation patterns. Experts believe the mid-Holocene warmed the Earth 1.7–3.3°F (1–2°C) warmer than it is today. Additional fieldwork has provided evidence that regions were much wetter during this period than they were during earlier arid conditions.

Late Holocene Neoglacial Fluctuations

This period was witness to a number of climate fluctuations spread over 4,500 years. Overall, experts say that temperatures were lower during this time than during the Mid-Holocene Thermal Maximum. From 2,500 to 4,500 years ago, a cooling period called the Iron Age neoglaciation occurred. The climate then heated around the time of the birth of the Roman Empire but cooled again during the Dark Ages (476–1000 C.E.). After this period, the climate warmed during what was referred to as the Medieval Optimum, or Medieval Warm Period. At this time, climate was sufficiently favorable to allow large-scale exploration and expansion of civilizations.

The Little Ice Age

The Earth's climate remained relatively warm until approximately 1450 C.E., then it took another turn toward cold. The subsequent cold period lasted from 1450 to 1890 and has become known as the Little Ice Age. The Little Ice Age was a time of renewed glacial advance and affected the North Atlantic, Europe, Asia, and North America. It alternated between colder and less cold periods. The two coldest time segments occurred in the 1600s and 1800s. The 1500s and 1700s were less cold. Because the Little Ice Age occurred in such recent times scientists were able to make scientific observations, take measurements, and keep historical records, enabling this event to be well documented.

According to the Intergovernmental Panel on Climate Change (IPCC), this episode was a modest cooling that mainly affected the Northern Hemisphere. Temperatures cooled about 1.7°F (1°C). The majority of the evidence of the period occurs in the Northern Hemisphere, although there are a few indications in the Southern Hemisphere that it, too, felt some of the effects.

There are several scientific opinions of when it actually began. Some claim it was the mid-1200s because that was when the Atlantic pack ice was first documented as beginning to advance south; others say the early 1300s because that marked the period when warm summer months were no longer dependable in Europe and farmers began to face extreme hardships growing food and supporting the populations that depended on them for their food supply. To support this, the period from 1315–17 is known as the Great Famine.

There were many effects from the Little Ice Age that were recorded and can be found in historical accounts today. Many written records discuss the extremely hard winters, especially those experienced in Europe and North America. As glaciers advanced in Switzerland, Germany, and Austria, farmers lost their farms, and entire villages were buried and destroyed by ice and snow. Holland, known for its internal waterway passages, became frozen over and unnavigable, as did others, including major waterways such as the River Thames in southern England. There are many accounts of people skating on the rivers and holding what they called frost fairs for recreation and entertainment.

The Little Ice Age also had serious ramifications for military, civil, and cultural issues. In 1794–95, the French army, under the leadership of General Jean Pichegru, overran Holland and captured the Dutch fleet with cavalry after their vessels became locked in the ice. Had the river not been frozen, history may have been different.

During the harsh winter of 1780, New York harbor froze over. Historical accounts say that people were able to walk across the ice from Staten Island to Manhattan. Business and commerce was devastated in Iceland during this same period because sea ice closed all the island harbors to ships, and they were not able to enter or leave ports. The hardships were so extreme that many people died. The same fate met the Vikings who had settled colonies in Greenland during the prosperous warm period prior to the Little Ice Age. Conditions became so harsh and cold in Greenland that the Norse could not farm the land, which caused them to die off. Winters in North America were also harsh, and records of numerous struggles were kept by the American Indians and colonists. They documented that winters were extremely cold and wet.

Throughout Europe and North America, traditional farming practices had to be drastically changed to accommodate the short, unpredictable growing seasons that had suddenly become the norm. Some of the extremely cold temperatures recorded on the eastern coast of North America are thought to have been the result of a change in the strength of the North Atlantic thermohaline circulation, a major ocean current responsible for bringing warmth from the equatorial regions to the polar regions.

Effects were also felt in other places worldwide. Ethiopia, for example, had permanent snow on mountain peaks where it no longer exists today; the Niger River in western Africa flooded (it has never flooded since); ocean sediment cores taken from Antarctica reflect relative ice age conditions during this period; it snowed in 1836 in Sydney, Australia, which is the only time since European occupation that has occurred; and tropical Pacific corals show indications of intense El Niño activity in the mid-1600s.

Experts, such as Gavin Schmidt, Drew Shindell, and David Rind of NASA GISS, have proposed a couple of major external forcings as the reasons for the cause of the Little Ice Age. They believe it may have been caused by a decrease in solar activity and an increase in volcanic activity. They have also proposed that a drastic decrease in population in Europe, Asia, and the Middle East because of famine and death coupled with a decrease in agriculture as well as the absorption of CO_2 through reforestation may have prolonged the Little Ice Age. Therefore, it probably was not due to one factor but to a combination of several global and regional internal and external factors.

The decrease in solar activity theory is supported by the known Maunder Minimum from 1645–1715. This was a well-documented period of low solar activity and happens to coincide with the coldest phase of the Little Ice Age. In addition, an increase in volcanic activity was documented during the same period, and repeated eruptions of sulfur and ash could have played a significant role in blocking incoming solar radiation. One well-documented example is the eruption of Tambora in Indonesia in 1815. The following year has become known in historical records as the Year Without a Summer, because frost and snow existed throughout July in the United States and northern Europe.

The Little Ice Age ended abruptly around the mid-1800s. Although some skeptics of present-day global warming suggest the Earth is still warming today in response to its recovery from the Little Ice Age, most

CLIMATE TIME LINE AT A GLANCE

Climatologists have been able to reconstruct paleoclimatic periods over the Earth's geologic past. The following list summarizes the major climate change episodes.

Ancient Climates—2 MYA to 4.6 BYA
Faint Young Sun Paradox
Snowball Earth
Hothouse Earth

The Pleistocene Climates—10,000 years to 2 MYA
Penultimate Interglacial Period
Dansgaard-Oeschger Events
Heinrich Events

Holocene/Recent Climate—10,000 years ago to today
Younger Dryas Cooling
Mid-Holocene Thermal Maximum
Late Holocene Neoglacial Fluctuations
The Little Ice Age

As scientists better understand the climatic events from the past, they are better able to apply their knowledge to the complexities of the climate system today and in the future in an effort to deal with the issues of global warming.

experts disagree, citing instead the fact that humans are causing an accelerated greenhouse effect through the burning of fossil fuels and other carbon-emitting activities.

KEY UNCERTAINTIES

The earliest-kept records of temperature measured officially by thermometers began in western Europe in the late 1600s. As more weather reporting stations came into existence and began recording official temperatures, by the early 1900s temperature networks had been established nearly worldwide. The exceptions were the polar regions. Collections began there in the 1940s and 1950s. Today, NOAA compiles the records from a climatology network of more than 7,000 stations worldwide so that studies can be made.

As this chapter has illustrated, the climate over the past several hundred thousand years has had a number of series of cold and warm periods. Today, the Earth is in a relatively calm period following the last ice age, but the temperature is slowly rising and at present-day levels passing earlier record high temperatures—this time due to human influence.

Although scientific understanding has progressed and dating techniques have improved, allowing scientists to increase their understanding about the Earth's past climates and how the climate system not only varies naturally but also responds to changes in climate forcing, key uncertainties still exist about the Earth's ancient climate. Even though scientists have a good understanding of the glacial-interglacial variations in climate and greenhouse gases, they recognize that there are still unanswered questions as to the exact factors that control these changes. According to the IPCC, climatologists need a better understanding of what causes abrupt climate change as well as a deeper knowledge of the main thresholds that can trigger, when they are crossed, rapid sea level rise and regional climate change.

Another area that needs improvement is climate modeling, particularly the simulation of abrupt changes in ocean circulation, flood frequency, drought cycles, monsoon behavior, the frequency of El Niño events, and the processes that control the advance and retreat of ice sheets.

Specific issues that present problems in recreating the past are that there are not enough paleoclimate records from the Earth's tropical areas, Southern Hemisphere, and oceans. Along with this, it would be beneficial to have data from areas evenly distributed over the Earth's surface, enabling climatologists to gain a more global view rather than a sporadic regional view, whereby large geographic areas must be interpolated. Experts believe if they could collect a larger global paleodatabase, they could compare the changes contained within it with the changes being seen today on a global scale and gain insights on how today's inputs are affecting the climate system in an unnatural way and how humans are upsetting the natural balance.

Geochronology and Climate Proxies

In order to determine when significant past climatic events occurred and changes shaped the Earth, climatologists must have a way to determine the geologic time period during which these events took place. They must be able to build an accurate climate time line. This chapter focuses on the various methods scientists use today to determine these specific intervals. It will look at both radiometric and nonradiogenic dating methods. It will also explore the importance of climate resolution and the dating techniques used as well as the concept of climate proxies, what they are, how they are used, and what they can reveal about past climate.

RADIOMETRIC DATING TECHNIQUES

When talking about dating objects and determining when a particular event took place or a specific climate existed on Earth, it is important to understand the difference between relative age and absolute age, because both are often referred to when discussing past climate.

Relative dating is simply determining if something is older or younger than something else. It does not provide an exact numerical age, only a comparable ranking. This follows what geologists refer to as the law of superposition, which simply means that when rock formations are formed, the oldest layers are on the bottom and the younger layers are above. Each layer of rock is younger than the layer it is sitting upon. The exception to this rule is if an area of existing normal-layered rock has been subjected to severe folding and faulting and its layers have been upended and overturned relative to each other.

When examining sedimentary layers, the relative age refers to the age of the deposition, not the actual date of the material in the formation, because they are not the same thing. Unlike igneous rock, which cools after formation and can be dated radiometrically, sedimentary rocks are composed of the weathered sediment of other parent material (other rocks). Therefore, the actual sediments may be much older than the sedimentary formation itself. These layers are given relative ages when they can be associated in the formation with other datable rocks, such as igneous rocks. For instance, if a layer of sedimentary rock is sandwiched between an igneous layer 50,000 years old and another 52,000 years old, then the sedimentary layer must be between 50,000 and 52,000 years old.

Absolute dating is the dating technique that allows ages to be assigned to samples in terms of the number of years before the present, based on a specific timescale. Absolute dating can be achieved in a number of ways. For example, it can be calculated based on natural annual cycles of trees, lakes, and glaciers; radioactive sequences; historical records; and trapped electrons.

Many types of physical evidence exist that can give climatologists clues as to what climate was like in the past. It can be seen in rock formations. If a layer of sedimentary rock exists in a presently arid landscape, it provides a clue that at one time there was an ocean or lake at the location. If a geologist can determine how old a rock formation is, then climate scientists will have a better understanding of what conditions were like at that site during that time. Likewise, if fossils from coral are found at polar latitudes, it tells scientists that at one time that location had been much warmer. If the plant remains of palm trees are found at a far northern latitude, it would indicate a warmer polar climate at one time. If the palm can be dated in a laboratory, then a date can be assigned to that warm period, increasing scientists' knowledge of the climate sequence.

One commonly used dating technique is radiometric dating, also called radioactive dating. In nature, rocks are made of many individual crystals, and each crystal is made up of different chemical elements, such as iron, magnesium, calcium, and sodium. All the Earth's known elements are listed in the periodic table (see p. 167). In nature, most of these elements are stable, meaning that they do not change over time. There are a few elements, however, that are not stable—they do change. In these particular elements, some of the atoms eventually transform from one element into a completely different element through a process called radioactive decay. The original element is referred to as the parent element. The element that the parent changes, or decays, into is called the daughter element. The aspect of this radioactive decay process that makes it such a useful tool is that the decay occurs at a predictable, known rate.

Radioactive decay works something like a clock or an hourglass. As soon as the clock begins ticking, the particular item can be analyzed in a lab and its age calculated. The decay occurs at a known rate, called the decay constant. This is the measurement of how much parent element decays to daughter element over a specific time period. It is this specific rate of decay that works like a clock. It is a special clock, however. Its time line is measured by a unit called a half-life.

One half-life is the amount of time it takes for half of the parent element to decay to the daughter element. This means that the first half-life reduces the parent element to half its original amount. The second half-life reduces the remaining parent element to half of that half (or one quarter of the original amount). The next half-life has half of that remaining amount decay over the same time interval (or one-eighth of the original amount), and so forth. The half-life time interval remains constant, but each time only half of the remaining parent material decays. By the tenth half-life interval, less than one-thousandth of the

THE DISCOVERY AND USE OF RADIOACTIVE DECAY

Natural radioactive decay was discovered by Henri Becquerel, a French physicist, in 1896. Shortly afterward, Ernest Rutherford, a British physicist, described the structure of an atom. These two discoveries are what prompted the idea of using radioactivity as a tool with which to measure geologic time. Then, in 1907, Professor B. B. Boltwood, a radiochemist at Yale University, published the first list of geologic ages of formations based on the use of radioactivity as a true laboratory dating process. His initial list of ages was not completely accurate, but it was still a significant breakthrough because he was moving in the right direction by measuring time in extremely large units of time: hundreds of thousands to millions of years.

Since these early achievements, the geologic timetable has been revised as technology has advanced. Scientists have made new discoveries, and techniques have become more sophisticated and accurate, allowing more of the geologic mysteries of the Earth to be unlocked. As technology continues to advance, so will scientists' understanding of the complex processes that shape the Earth.

original number of radioactive parent atoms is left. Because this clock is so consistent—it cannot be influenced or changed by external forces, such as heat, cold, acceleration, pressure, vacuum, or chemical reactions—it has proven to be a reliable way to date materials in the Earth's crust and on the surface.

An event must occur to start these radiometric clocks ticking. If it is rock that is being dated, igneous rock is the most commonly used because it cools very quickly from its molten lava state. As soon as the lava cools to the point that neither parent nor daughter elements can leave or enter the rock, it forms what is called a closed system—nothing leaves or enters; the system stays as it is. Once a system is closed, it cannot be influenced by outside forces; it can be influenced only by internal forces, in this case, internal radioactive decay of the parent element. As the parent exponentially decays ($\frac{1}{2}$ to $\frac{1}{4}$ to $\frac{1}{8}$ . . .) the daughter

exponentially increases in abundance. Radioactive isotopes are useful for the first five or six half-lives. After that, there is not enough parent material left to be reliable.

Most radioactive isotopes have rapid rates of decay—or short half-lives—and lose their radioactive properties within just a few days or years, which prevents them from being useful dating tools. There are some isotopes that decay slowly, however, and this allows them to function as extremely useful tools for determining age. Radiometric dating is a versatile dating tool. Different types of radioactive parents have different half-lives, so depending on the age range of an item being dated, various parent elements are used for different parts of the Earth's history. Of course, the parent element must be present in the rock being dated. If the oldest rocks on Earth are being dated, for instance, the uranium-to-lead series is used because it has a very long, slow decay rate.

Occasionally, there can be problems with the dating technique. It is assumed that when the clock is set there is no daughter present, but that is typically not the case, which adds some error—the sample is usually not pure. Also, occasionally, the system is not completely closed, allowing parent or daughter isotopes to leave the system, which can skew the results.

Today more than 40 different radiometric dating techniques are used that are each based on a different radioactive isotope. Isotopes are forms of a chemical element that have the same atomic number but differ in mass. Isotopes with long half-lives decay very slowly, which makes them useful for dating ancient events. The isotopes that have shorter half-lives are not used to date ancient events because there would not be enough parent isotope left to measure. Isotopes with short half-lives are useful to date events over shorter intervals of time. The half-lives are measured by a radiation detector in a lab or by measuring the ratio of daughter to parent atoms in a sample.

The table on page 47 shows some of the most commonly used radiometric techniques in climate science.

When dating climate records, obtaining the date based on radiometric techniques is the first step. Once dates are obtained on igneous rocks, they serve as markers when looking at the geological formations and give boundaries on intervals of time as to when sedimentary layers

Radioactive Decay Used to Date Climate Records				
PARENT ISOTOPE	DAUGHTER ISOTOPE	USED ON HALF-LIFE	USE ON AGE RANGES	MATERIALS
Uranium-238	Lead-206	4.5 Byr	>100 Myr	Multiple rocks
Uranium-235	Lead-207	704 Myr	>100 Myr	Multiple rocks
Rubidium-87	Strontium-87	48.8 Byr	100 Myr	Granite
Thorium-230	Radon-226*	75,000 years	<400,000 years	Corals
Potassium-40	Argon-40	1.25 Byr	>100,000 years	Basalts
Carbon-14	Nitrogen-14*	5,780 years	<50,000 years	Anything containing carbon

(Notes: * Daughter is an escaped gas that cannot be measured; Byr = Billion years; Myr = Million years)
Source: U.S. Geological Survey

were formed. It is within these sedimentary layers that many clues to the Earth's past climate can be found.

Most sedimentary rocks, such as sandstone, shale, and limestone, are related to the radiometric time scale only by bracketing them to datable igneous rock formations. This way sediments can be dated based on their relative positions to the datable igneous formations. In other words, once a radiometric age is obtained for an igneous formation, then the ages of the sedimentary layers are constrained by their positions above, below, or between igneous layers. There are six radiometric dating techniques commonly used in climatology: (1) uranium-lead, (2) rubidium-strontium, (3) dating using thorium-230, (4) methods using lead, (5) potassium-argon, and (6) carbon-nitrogen.

In cases in which there are not enough igneous rocks present to serve as markers, fossils are commonly used. The best types of fossils are those species that did not exist for long on Earth but are found in many places worldwide, making their possible temporal range naturally confined. Areas that have fossil records that record the extinction of one species and the appearance of new species are also good for establishing sound geologic time markers.

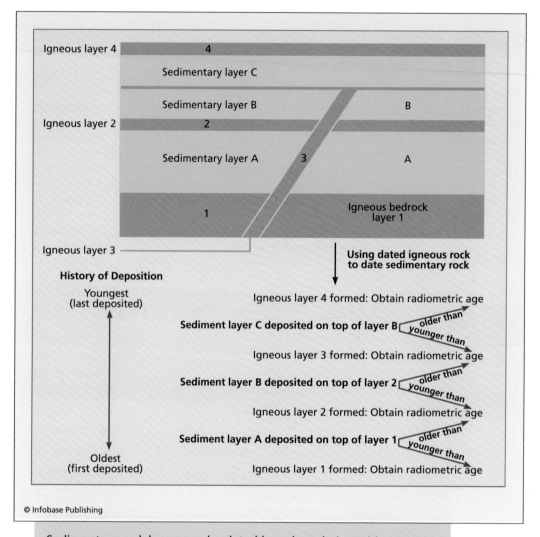

Sedimentary rock layers can be dated based on their position to igneous rock formations in the geologic column. Once the igneous rocks are given absolute dates through radioactive dating, the sedimentary rocks are given ages constrained by their positions above, below, or between the igneous layers.

Uranium-Lead

The uranium-lead technique has been used the longest; it was introduced in 1907. There is more than one useful uranium-lead dating sequence. Natural uranium consists of two isotopes, U-235 and U-238.

As shown in the chart on page 47, these two isotopes decay at different rates to produce lead-207 and lead-206, respectively. Because of this, it is possible to get two age estimates for each sample. Because U-235 has a half-life of 704 million years and U-238 has a half-life of 4.5 billion years, each sample can be cross-checked, ensuring greater accuracy. Sometimes, however, it is hard to keep uranium and lead in many minerals in which they are found. Because of this, the uranium-lead dating technique can be less reliable than other measures.

Rubidium-Strontium

Rubidium has a half-life of 48.8 billion years. This method is used on old igneous and metamorphic rocks. Sometimes, there can be confusion with this method if a rock sample contains some minerals that are older than the bulk of the rock. This may occur during formation if the rock accidentally picks up an unmelted mineral from the surrounding rock that the magma passes through. Problems can also arise if the rock has undergone some metamorphism. In this case, if some parts of the rock completely melted and others did not, it will confuse the sampling equipment as it tries to read it as either old igneous or younger metamorphic rock.

Dating Using Thorium-230

Dating techniques using thorium are employed on oceanic sediments that are older than the range that radiocarbon techniques can detect. Uranium that is present in seawater eventually decays to thorium-230 (called ionium), is precipitated into the ocean floor sediments, and is stored. Thorium-230, which is part of the uranium-238 decay series, has a half-life of 80,000 years. Protactinium-231, which is derived from uranium-235, has a half-life of 34,400 years. Both can be used for age calculations. This method has been used on deep-sea sediments formed during the past 300,000 years.

Methods Using Lead

Methods involving lead can be applied to rocks younger than Precambrian (542 MYA). In the uranium-lead method, the age of geologic material is calculated based on the known radioactive decay rate of uranium-238 to lead-206 and/or uranium-235 to lead-207. In addition, if the thorium-232 to lead-208 series is also applied to a sample, then three independent ages can be calculated for an object for validation. Geologists use this method most often on Precambrian materials.

Potassium-Argon

Potassium-argon is one of the simplest dating methods and has a half-life of 1.3 billion years. It can be used on rocks within a wide range of ages—a few thousand years to billions of years. Potassium is present in most rock-forming minerals, and in samples that have Potassium-40 in them, there is generally enough Argon-40 present that the rock can be accurately dated, even if there is only a small amount of Argon-40 present. Potassium is a common element found in clay minerals, tephra, micas, and evaporates. Ages can be measured very accurately with this technique. Usually, when geologists date rocks, they run more than one method of analysis if possible in order to confirm the integrity of the results.

Carbon-Nitrogen

Pertinent to younger geologic formations, a radiometric method that is often used is one that specifically dates carbon-bearing sediments. Carbon-14 is continuously produced in the Earth's upper atmosphere through the bombardment of nitrogen by neutrons from cosmic rays. The process creates radiocarbon, which then becomes uniformly mixed with the nonradioactive carbon in the atmosphere. This process differs from other methods because it dates the carbon-bearing pieces of evidence directly rather than dating the rock in which it is found and inferring an age. The technique called "radiocarbon dating" has been used extensively since the 1950s.

All plants that live on the Earth extract carbon from the atmosphere for photosynthesis. Animals acquire it from the consumption of plants and other animals. A small portion of that carbon—carbon-14—is radioactive. When the organism dies, it stops exchanging carbon with the atmosphere and starts the decay process. In other words, the clock begins ticking. The carbon-14 parent decays to the nitrogen-14 daughter, which is a gas that escapes. The amount of carbon-14 loss is mea-

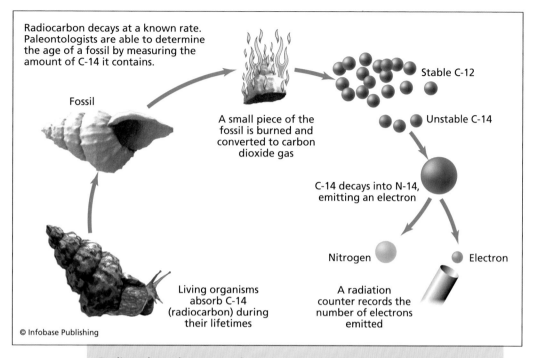

Radiocarbon decays at a known rate. Paleontologists are able to determine the age of a fossil by measuring the amount of C-14 it contains.

Fossil

A small piece of the fossil is burned and converted to carbon dioxide gas

Stable C-12

Unstable C-14

C-14 decays into N-14, emitting an electron

Nitrogen

Electron

Living organisms absorb C-14 (radiocarbon) during their lifetimes

A radiation counter records the number of electrons emitted

© Infobase Publishing

Radiocarbon decays at a known rate, making it possible to date objects containing carbon. Once an object dies, it stops absorbing carbon-14. The carbon-14 becomes unstable and decays into nitrogen-14, where it can be measured and its age calculated.

sured by comparing how much carbon-14 there is compared to how much stable isotope of carbon-12, which is not removed by radioactive decay, there is.

Carbon cannot be used to date extremely old objects or events, because its half-life is only 5,730 years. The reasonable limit for which carbon dating can be expected to be reliable is about 30,000 to 50,000 years. Any type of material that contains carbon can be carbon dated, including anything that lived, such as wood, twigs, leaves, charcoal, and bones. Many samples used for carbon dating come from lake environments, which are poor in oxygen. Because of the lack of decay, such samples are well preserved.

One thing that must be kept in mind with any of the radiometric dating techniques is that their precision depends on how carefully the procedure is performed. If a sample being dated becomes contaminated

Isotope Techniques	
Uranium to lead (minerals)	1 million to 4.5 billion years
Rubidium to strontium (minerals)	60 million to 4.5 billion years
Potassium to argon (minerals)	10,000 to 3 billion years
Uranium series disequilibrium (minerals, shell, bone, teeth, coral)	0 to 400,000 years
Carbon-14 (minerals, shell, wood, bone, teeth,water)	0 to 40,000 years
Radiation Exposure Techniques	
Fission track (minerals, natural glass)	500,000 to 1 billion years
Thermoluminescence (minerals, natural glass) and optically stimulated luminescence (minerals)	0 to 500,000 years
Electron spin resonance (minerals, tooth enamel, shell, coral)	1,000 to 1 million years
Other Techniques	
Geomagnetic polarity timescale (minerals)	780,000 to 200 million years
Amino acid racemization (shells, other biocarbonates)	500 to 300,000 years
Obsidian hydration (natural glass)	500 to 200,000 years
Dendrochronology (tree rings)	0 to 12,000 years
Lichenometry (lichens)	100 to 9,000 years

Axis (Years ago): 0   10   100   1,000   10,000   100,000   1,000,000   10,000,000   100,000,000   1,000,000,000

Years ago

© Infobase Publishing

Different radiometric dating methods are only useful over specific time periods. The proper method must be used for the right sample in order to obtain reliable results.

with a material of another age, it can give a false result. The sample can also become naturally contaminated if isotopes mix with the sample after its original creation. Also, it is important that the right technique (half-life) for the right sample type be used.

According to the U.S. Geological Survey (USGS), thousands of materials that have been dated by radiometric techniques are available worldwide and can be used to bracket geologic formations in order to discover more about the Earth's geologic past. As more information is collected and more formations are dated, geologists gain a better understanding of radiometric dating techniques. As they gain more knowledge, the geologic timescale becomes more refined. At present, the geologic timetable represents the current state of knowledge, but as discoveries are made, the table is revised and modified to reflect new information about the Earth's ancient past.

NONRADIOGENIC DATING METHODS

Other dating techniques are available that do not use the radiometric properties of materials. These techniques are often used on objects dating from the past 100,000 years of the Earth's history and involve correlating physical phenomena with natural cycles. The most well-known technique is dendrochronology, also known as tree-ring dating. Other cyclical techniques include analyzing ice cores and varves (annual sediment layers from lakes). These methods will be discussed in greater detail in chapters 5 and 6. Three nonradiogenic dating techniques climatologists rely on are (1) orbital tuning techniques, (2) thermoluminescence, and (3) electron spin resonance.

Orbital Tuning

Orbital tuning is an important dating technique climatologists use to study past climate. This refers to a procedure of linking cycles of incoming solar radiation with the Earth's ice volume responses to determine age. As discussed earlier regarding the Milankovitch cycles, astronomers have calculated the solar radiation signal over time and have determined that cycles at 41,000 and 23,000 years have produced regular climatic responses. Because of this correlation, these predictable climate cycles can be used as time clocks for dating purposes. Scientists have also determined that ice volume response in the Earth's ice sheets always lags behind these orbital forcings by a consistent amount, allowing them to date climate records in ocean sediments in relation to the known timing of the orbital changes. They are able to correlate this through the detection and analysis of oxygen-18, because $\delta^{18}O$ is a good index of ice volume. Every 32.8 feet (10 m) of change in global sea level causes a 0.1 percent change in $\delta^{18}O$. Scientists can also date the reversals of Earth's magnetic field in ocean sediment cores to add additional absolute dates to a profile. In fact, many climate experts regard the timescales derived from orbital tuning chronometry for the last several million years to be more accurate than those based solely on radiometric dating techniques.

It is also a versatile technique—it can be applied to any ocean sediment core containing $\delta^{18}O$. When this timescale is established in the higher (polar) latitudes, it is often correlated at lower latitudes to reconstruct orbital-scale monsoon activity at 23,000-year cycles as well as decay of ice-sheets at both middle and high latitudes. Climate change back to 5 million years has been established using this technique.

Thermoluminescence

Thermoluminescence is another dating technique. It does not rely directly on half-lives but uses the fact that radioactive decay causes some electrons to end up in a state of higher energy. The longer the radioactivity, the more high-energy electrons collect. If the material is then taken into a lab and heated, it puts the higher-energy electrons back into their normal orbits. As the orbits return to normal, a small amount of light is given off. This measurement can be calculated into an age. This technique is useful to date materials less than half a million years old.

Electron spin resonance

Electron spin resonance, also called electron paramagnetic resonance, also works because of changes in electron orbits caused by radioactivity over a time interval. This method can be used for time periods up to 2 million years and is mostly used on carbonate materials such as coral reefs and cave deposits.

CLIMATE RECONSTRUCTION RESOLUTION

When past climate is reconstructed, it is important to keep several factors in mind. When the climate record is preserved, integrity depends on whether a site is disturbed or left relatively untouched. Another important factor is how fast the record is accumulated. This determines how protected the site is from disturbance. If a site is left exposed for a long period of time, chances of it being contaminated or damaged are greater.

When climate data is retrieved from sedimentary rocks, it is generally in areas that have been in low-energy environments for a long period of time, rather than turbulent areas that have had consistent

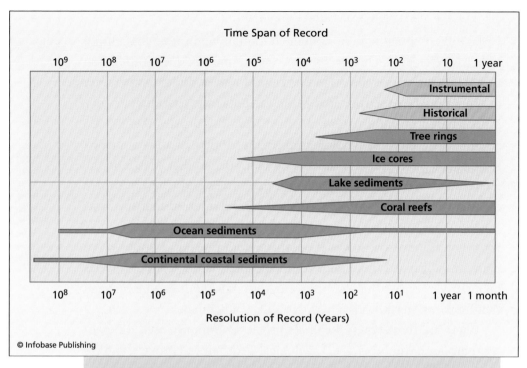

Time Span of Record

10^9 10^8 10^7 10^6 10^5 10^4 10^3 10^2 10 1 year

Instrumental

Historical

Tree rings

Ice cores

Lake sediments

Coral reefs

Ocean sediments

Continental coastal sediments

10^8 10^7 10^6 10^5 10^4 10^3 10^2 10^1 1 year 1 month

Resolution of Record (Years)

© Infobase Publishing

Different climate markers are applicable to specific time periods. More recent time spans are dated by tree rings, historical records, and instrumental records whereas older time periods are dated by continental and ocean sediments.

wave action. Coastal marine areas can have 3 feet (1 m) of deposition of sediments per year; in lakes a few thousandths of an inch (millimeters) per year; and in deep-sea environments a few thousandths of an inch (millimeters) per thousand years. The amount of disturbance by animal life varies per area. Coastal regions have more activity than do deep-sea and lake environments. Generally, lake sediments have the best resolution and deep-ocean sediments the worst.

Each type of archival record is geared toward a specific time period. Therefore, what climate record is being dated, where it is geographically located, and what it is composed of directly determine which geochronological method is used to date it in order to interpret past climate. As the figure shows, additional dating techniques exist for the more recent past.

CLIMATE PROXIES

Clues to past climate can be contained in many indicators that climatologists refer to as climate proxies. (*Proxy* means "substitute.") The term is used because climatologists cannot obtain direct temperature or other climatic data from proxies but can only infer past climatic conditions based on information obtained from the proxies themselves. As an illustration, it is possible to infer the climate of an area just by looking at a photograph. If the photograph, such as that shown on the facing page, is of a beach dotted with palm trees and tropical flowers along an ocean front with blue skies, it can be inferred that the location is a tropical setting somewhere near the equator, because we know those conditions exist today. If another photo portrays a snow-swept plain, snow-capped mountains in the distance, and a team of huskies being driven across the plain by a fur-clad musher, it can be inferred that the location shown is a high-latitude, cold climate.

There are three types of climate proxies commonly used: (1) geomorphic landform proxies, (2) geological-geochemical proxies, and (3) biotic proxies. Geomorphic landform proxies involve identifying features on the Earth's surface indicative of specific climatic events, such as glacial landforms, landforms caused by desert environments, landforms caused by running water, and those that result from wind. Geological-geochemical proxies involve movements of the Earth's materials through the climate system either as physical particles (such as sediments) or dissolved chemicals. Examples include sediments deposited by water, ice, and wind in soil profiles; rocks; varves; glacial ice; loess deposits; paleolimnology; and cave formations. Chemical weathering processes involve dissolution (the dissolving of carbonate and evaporite rocks) and hydrolysis (the addition of water to rocks during weathering). Biotic proxies include evidence obtained from the fossil data of plants and animals, including pollen, plankton, cones, seeds, leaves; corals; insects; paleofire; and tree rings. These proxies will be described in greater detail in the following three chapters.

Another type of proxy data is historical data—a written record. These documents can contain a wealth of information about climates of the past. Over time, people have kept records in all kinds of settings:

South shore of the island of Kauai, Hawaii *(Nature's Images)*

farm records, ship logs, explorer's accounts, travelers' diaries, newspaper stories, civic records, and personal accounts. Many sources of verified historical data have augmented the instrumental records of the past 150 years, further extending direct knowledge of the climate backward in time.

Proxy Data:
Geomorphic Landforms

Climate has left its mark in several places on the planet—in the chemical and physical structures of the land, the oceans, and life. These climate artifacts, called climate proxies, reveal climate patterns that can extend backward in time hundreds of thousands, even millions, of years. When this proxy evidence is combined with present-day observations of the Earth's climate and is entered into computer models, this paleoclimatic data (*paleo* means "ancient") can help scientists predict future climate change. This chapter illustrates how climatologists can reconstruct a clearer picture of past climates by unraveling the stories of the various landforms on the Earth's surface. The Earth's geologic record represents a rich treasure trove of information reflecting the ways in which climate has left its lasting mark on the Earth's geographic features.

Evidence is reflected in various landforms and landscapes detailing the Earth's past temperature, moisture, humidity, and atmospheric circulation patterns. These climate proxies are in the form of specific geo-morphic landforms. Geomorphology is the study of landforms and the various processes that create and shape them. These resultant landforms are the result of climatic variables such as the mean annual, maximum and minimum, and seasonal distribution of temperature, net evaporation, and levels of precipitation on land and currents, salinity, nutrient level, and temperature in the oceans.

When scientists reconstruct past climate, they do it in a methodical way. The first phase consists of collecting the actual proxy data. Once collected, they are analyzed, measured, and then calibrated (adjusted to fit) with present-day climate records. When climatologists calibrate data, they assume that the laws of ancient climate behaved in a similar way to modern climate. They are aware, however, that past climates could have behaved somewhat differently and that some error could be introduced during the reconstruction, but they do their best to ensure the data are interpreted as accurately as possible. The data from the ancient climate are then compared to present climate so they can be interpreted.

According to the U.S. Geological Survey's (USGS) Geologic Division, climates are always changing, and understanding these changes and their effects are some of the most challenging issues society has to deal with today. This is especially the case with global warming. They point out that it is a major issue facing society today and deserves a high degree of attention in the form of research because many unanswered questions remain. One of the biggest uncertainties is the primary driving source—how human activity affects the Earth's natural cycles. Other uncertainties include the triggers of rapid climate change and how people can plan for, adjust to, and protect themselves from the climate if it does change with little warning.

Again, it comes to understanding the climate's past in order to make sense of it and predict its future. Looking at the Earth's geologic and geomorphic past is one way to do this. The forces on the surface of the Earth that shaped its features can give climate scientists significant clues about what the climate was like on Earth in ancient times.

Geologists know that climate has a major influence on the shape of the landscape. Different factors work on the landscape over time, shaping it in unique ways, such as water, wind, energy, gravity, and temperature. A desert landscape, for example, is very distinct. If relics of arid land features, such as dunes, alluvial fans, pediments, desert pavements, and scoured and eroded sedimentary surfaces, appear in a humid climate, this is conclusive evidence that the area was arid at some time in the past. Likewise, glacial features are distinct, and their presence writes another specific story about what past climate was once like.

GEOLOGIC EVIDENCE

The geologic processes that evolve most abruptly are those that dominate and determine the appearance of a landscape. Factors such as climate, rock type, steepness of terrain, presence of water, and presence of wind are all contributing factors to a landscape's ultimate appearance. When landscapes are young and are being actively uplifted, the gradient is steep. When water, such as from a river, begins to erode under the force of gravity, it moves downslope with great erosive power. Young geologic features reflect these forces with their angular features.

In direct contrast, when a landscape has reached old age, the eroded features become level and rounded. This type of evidence—existing drainage channels and degree of erosion—tells geologists the history of the landscape in relation to mountain-building processes and the abundance of a relevant water source, both variables that also indicate what the past climate was like.

The same concept applies to the geologic evidence of weathering. In dry, arid climates, the weathering process is very slow; in humid climates, it is much faster. Therefore, the degree of weathering over a determined time period gives Earth scientists clues as to what the past climate was like. This is why arid landforms contain many angular features, such as pinnacles, outcrops, and scree (angular boulders). In warmer, more humid climates, such as tropical, chemical weathering processes dominate, leaving distinctive landforms. In extremely cold areas, still other features are formed, characteristic of the annual freeze/thaw cycles, clueing the paleoclimatologist in to ancient conditions. The same holds for glacial landforms. They are very distinctive, and when seen in the field—even if it is in the Tropics today—they tell the experts that the area was once cold enough to provide the right environment for a glacier. Evidence of oceans and seas once existing in areas that are dry and mountainous today are indicative of a very different climate in the past. Landscapes around the world speak of climates that existed long ago, and when geologists can determine their age, it makes it possible for climate scientists to reconstruct past climate events.

LANDFORMS OF ARID ENVIRONMENTS

Water has enormous erosive potential in arid environments. Changes in climate causes stream terraces to form. Stream terraces are formed when streams carve downward into floodplains, leaving steplike benches along the sides of a valley. In present-day landscapes, when geologists find older stream terraced surfaces that have no obvious connection to a modern drainage system, it provides a clue to the area's past climate. Sometimes old terraces have been weathered in such a way as to expose some of the soil or subsurface structure for more insight into what the ancient ecosystem was once like. Stream terraces exist throughout the western regions of the United States, such as in the Mojave region of California and the San Juan River area of Utah.

When an area is cool and wet, there is ample vegetation growing on the landscape to protect the soil from erosion. During dry periods, however, plants struggle to survive, and many die off, leaving the ground bare and exposed to intense erosion during the infrequent but often powerful storms. Under the cooler, wetter conditions of an ice age, soil development and weathering processes are much more rapid. During multiple freeze-thaw cycles, when material is continuously wetted and dried, rocks are physically torn apart. In addition, biological activity, such as plant's roots penetrating the ground, can also weather rock, forming soil horizons. This weathering supplies large deposits of sediments that can be transported to new locations when a water source becomes active again. Climate is also a major factor in the formation of caliche, a calcium carbonate–rich crust that forms in the stony soil of arid regions. It forms when water evaporates at the ground surface and calcite cements the surrounding sediments together. Caliche is commonly found in the southwestern portion of the United States. Sometimes caliche deposits serve as a resistant caprock above isolated stream terraces, protecting them from erosion.

Another feature of arid landscapes that speaks of past climate are remnants of ancient lakes. In the United States, the last major ice age, which ended roughly 14,000 years ago, had a significant effect on the weathering and erosion patterns of the Southwest. In California and Nevada, many ancient lakes existed, such as Lake Tecopa, Silver Lake, Soda Lake (Lake Mojave), Coyote Lake, Lake Manley, Panamint Lake, Owens Lake, China Lake, and Searles Lake. These lakes were contained geographically within the area bordered today by the Sierra Nevada, San Gabriel Mountains, and San Bernardino Mountains in California and Las Vegas and the Colorado River in Nevada. Excavations into dry lake beds have produced fossils of shelled invertebrates, fish, and plants that could have survived only in a lake environment, allowing climatologists to piece together a climate time line. The climate-induced formation and disappearance of lakes has also influenced the development of river systems in the region, leaving evidence across the arid landscape. During the ice ages and wetter periods, large river systems probably existed with diverse fauna and flora.

One of the more famous of such lakes during this time was Lake Bonneville. It existed from about 32,000 to 14,000 years ago and occupied the lowest closed depression in the eastern Great Basin. At its largest extent it covered about 20,000 square miles (51,800 km²) of western Utah and smaller portions of eastern Nevada and southern Idaho. Melting glaciers fed the lake.

Today, researchers study the fossil record left behind by mammoths, mastodon, musk oxen, large-horn bison, camels, giant beaver, giant wolves, giant bears, big-horn sheep, giant ground sloths, short-faced bears, and saber-toothed cats. They also study the glacial deposits and landforms of the area as well as the distinct shoreline left by the lake, as seen in the photo.

Climatologists have identified several climate forcings in arid environments that help them successfully reconstruct the Earth's past climates. The arid ecosystem is so fragile and responsive that climate affects nearly all aspects of the biosphere and landforms in such a region. According to the USGS, the illustration depicts climate-induced forcings involving lakes, plants, animals, soils, and landforms.

During wet periods, the increased moisture allows lakes to form, promotes plant growth, allows species to expand their ranges, and

Midway up the mountain, the horizontal line is the ancient shoreline of Lake Bonneville, a major lake in the western United States during the last major glaciation 14,000 years ago. This shoreline represents direct evidence of past climatic conditions. *(Nature's Images)*

accelerates soil formation. During this phase of the cycle, sand supply to dune fields in arid regions is at a minimum. Then, when dry conditions prevail, the phase shifts: Plant cover decreases, and desert storms erode soil and provide a source of sand for alluvial fans. Scientists acknowledge that more research is needed to quantify both the short- and long-term effects of climate change on the various surface processes that affect the life-forms, water supply, and geomorphic processes that operate on the landscape, such as the effects of landslides, changing water tables, changing water quality, and erosion rates.

Weathering and erosion are continual processes in desert landscapes. The mechanical breakdown (physical weathering) of rock is

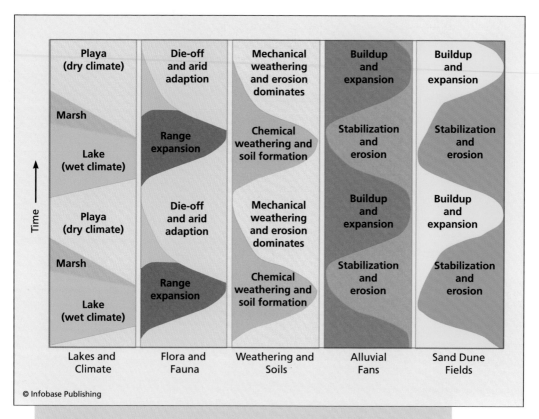

Lakes and Climate	Flora and Fauna	Weathering and Soils	Alluvial Fans	Sand Dune Fields
Playa (dry climate)	Die-off and arid adaption	Mechanical weathering and erosion dominates	Buildup and expansion	Buildup and expansion
Marsh	Range expansion	Chemical weathering and soil formation	Stabilization and erosion	Stabilization and erosion
Lake (wet climate)				
Playa (dry climate)	Die-off and arid adaption	Mechanical weathering and erosion dominates	Buildup and expansion	Buildup and expansion
Marsh	Range expansion	Chemical weathering and soil formation	Stabilization and erosion	Stabilization and erosion
Lake (wet climate)				

© Infobase Publishing

In a desert environment, climate is interrelated with many physical features, including lakes, plants, animals, soils, sand dunes, and other weathered material. Over time, wet and dry climates are each associated with specific desert features.

fairly rapid in arid climates. Another indicator of past dry climate is that soils form very slowly, leaving most bedrock exposed to erosion. There are many physical processes always at work on the desert landscape. Deserts have wide temperature ranges. Although it can become very hot during the day, it gets very cold at night. The constant cycle of heating and cooling weakens rocks and breaks them apart over time along areas of weakness, such as cracks. Once the cracks appear, plant roots can grow in them and widen them, and water can run into them. When water freezes, it expands; when it melts, it contracts and flows out. This is referred to as a freeze-thaw cycle. A rock can undergo hundreds of freeze-thaw cycles, which further weakens the rock and breaks it apart.

Checkerboard Mesa, Zion National Park, Utah *(Nature's Images)*

Exposure to wind and precipitation also causes distinct weathering patterns on landscapes that can clue geologists in to desert landscapes of the past. The geologic formation in the photo is Checkerboard Mesa, located at Zion National Park in Utah. The deep furrows in this formation have been cut like a checkerboard over millions of years. The horizontal furrows were carved out during the Jurassic period by windblown sand off huge ancient sand dunes that spread across the landscape. The furrows were then widened and split vertically in joints due to a climate favorable to multiple freeze-thaw incidents. This unique landform tells a story in itself about past climatic conditions of the area.

When rainfall occurs in desert regions, mountain slopes become saturated. This can cause what geomorphologists refer to as mass movement—the downslope movement of a portion of the land's surface. Mass movement can occur sporadically, as in the case of rock falls; quickly, as in the case of landslides and mud slides; or slowly, as in the case of creep. Although rain does not fall often in a desert, when a storm does occur, it is usually very destructive. Because so little vegetation is available to provide stability to the soil, the water is able to erode and move large amounts of weathered material in a single incident. Sometimes a huge boulder may end up on a flat desert plain as a result. Flash floods and debris flows leave their marks on the landscape, which can also be interpreted by climatologists.

Carbonate sedimentary rocks are another key arid environment feature. These are common in the southwestern United States. Carbonate rocks originate from limey sediments composed of the calcareous skeletal remains of algae and invertebrate shell material or from calcium that precipitates directly from agitated, warm seawater. This usually signals an environment such as a shallow continental shelf in a warm, tropical climate. These deposits alone give climatologists a good look at the climatic past of the American Southwest. Many of these deposits have been dated to the Proterozoic and Paleozoic Ages.

Most of the ancient limestone formed from planktonic algae, but in the late Paleozoic time, coral reefs also became key producers of carbonate sediments. Limestone is mostly made of calcite ($CaCO_3$). The Mojave Desert has many such deposits, suggesting its past climate was once a shallow ocean maritime environment. Freshwater limestone deposits referred to as tufa occur around springs and in areas where ancient lakes' wave zones reached them.

Pediments, alluvial fans, and playas tell their own stories about past climate. A pediment is a gently sloping erosion surface or plain of low relief formed by running water in an arid region at the base of a receding mountain front. An alluvial fan is a deposit built from material deposited from a stream that flows down a canyon and empties out onto a valley floor. Once the stream reaches the open valley and is no longer confined, it can migrate back and forth, depositing its sediment load in the shape of a large fan, with its point at the mouth of the canyon.

According to the USGS, the impact of climate change on alluvial fans has been the focus of recent research. They have conducted studies on alluvial fan deposits in Death Valley and throughout the southwestern United States that show that a period of increased alluvial fan deposition occurred between the last Glacial Maximum (15,000 years ago) and the beginning of the arid conditions of the early Holocene (9,400 years ago). In trying to determine whether it was climate change or active tectonics that was responsible for the creation of the fans, they

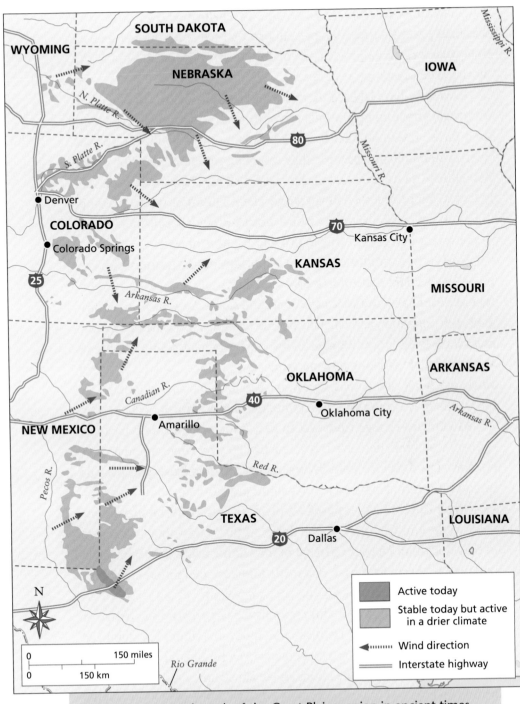

Sand dunes covered much of the Great Plains region in ancient times. Today they are stable, but if the climate becomes drier, as with increased global warming, the sand dunes could become active once again.

concluded that the climate of the region transformed over time from seasonably wet to arid. This period also marked a significant span of time when vegetation made a major shift from widespread flora to the more sparse vegetation seen today.

Playas can have lakes in them during wet periods. The dry lake bed is typically made up of stratified clay, silt, sand, and soluble salts. Common in the intermountain basins of the American Southwest, playas interest climatologists because they can often reveal information about past climate. During the last glacial period, many playas were lakes and marshes. Then, about 8,000 years ago, these bodies of water dried up. Today, they flood only after seasonal storms provide flash-flood waters or if a spring develops nearby to provide a source of water.

Sand dunes and dust are other important desert features that can reveal much about past climate. Sand and dust transport plays an important role in shaping the landscape. Dust and sand can come from weathered rock, alluvium, and dry lake beds. Prevailing winds can carry clay and silt particles large distances. Wind moves sand along the surface as a saltating bed load. The moving sand stalls and accumulates as dunes where the wind rises over a significant barrier, such as a mountain. As long as there is a source of sand, dunes develop and form.

As climates change from wet to dry, dune systems are repeatedly formed, stabilized, destroyed, and reactivated over time. Ancient dune systems exist today, covered by vegetation and stabilized, and provide important information to scientists about past climate. Some may even have fossil ripple marks exposed on them. Examples of stabilized dune fields in the United States are the dunes in the Plains states. As shown in the illustration on page 67, most of these dune fields are stable, but at one time in the ancient past, under the influence of a different climate, they were all very active.

According to the USGS, this region is the largest sand sea, active or stabilized, in the Western Hemisphere. Today, the dunes are stabilized by vegetation and do not pose a problem, even though the winds on the Great Plains are some of the strongest—even stronger than those of the world's deserts. USGS geologists determined that the sand dunes on the Great Plains were active within the past 3,000 years. Many were active during the 1800s, and some were even active during the dust bowl of the

1930s. This causes concern for climate scientists because the dunes have been active under climatic conditions that are not much different than those of today. If the dunes were reactivated because of global warming or natural climatic variation, it would cause significant changes to the land. For example, farm and range land could be destroyed, transportation networks shut down, and wetlands and wildlife habitats damaged. This is one major reason why the USGS conducts field investigations and works with land management planners.

LANDFORMS OF COLD ENVIRONMENTS

Cold climates are known for many distinct types of landforms. These landforms are associated with freezing and subfreezing temperatures along with the presence of water. Cold climate processes are often linked to frost action. During the last major ice age, ice covered nearly one-third of the Earth's surface. The Northern Hemisphere was buried under a massive ice sheet up to 2 miles (3.2 km) thick in places. The ice extended from the North Pole southward to southern Illinois. Greenland was buried under ice, as was much of northern Asia and northern Europe. As seen in earlier chapters, this was not the only time this happened. The Earth has been subjected to multiple advances and retreats of ice sheets throughout its history. Each ice age that occurs leaves its unique mark on the geology of the landscape. Landforms of cold climates are very distinct and, as such, can be identified and dated. They have proven to be one of the prime tools that has allowed climatologists to be able to piece together the climatic history of the Earth.

Even when glaciers are not involved, cold temperatures act on the ground. Repeated freezing and thawing of the ground acts to move and sort earth. If there is a slope present, it causes the soil to begin to slowly flow, or creep, downhill. This is a process called solifluxion. This churning activity within the ground causes the larger rocks to rise to the surface, while the smaller particles settle into the open gaps beneath that are created after the freezing action has caused the soil to expand. Giant wedges of ice can form in the ground. Over time, as this process continues, it creates an odd, distinct pattern on the ground when seen from above. It creates a polygonal patterning of wedges and circles of stones called patterned ground. When geologists see this, they know a cold climate existed there at one time. If the ground is in the tundra and drainage is poor, this process eventually forms a series of hills called pingos. Pingos can reach 98 feet (30 m) high.

There are several distinct glacial landforms that climatologists use to identify past glaciation. Glacial moraines are formed from the deposition of material eroded by a glacier. Where rock, soil, debris, and other matter is scoured off canyon walls and carried along, the material builds up as the glacial ice carries the heavy load like a conveyor belt. As soon as the ice melts, huge sinuous deposits are left along the glacier's sides and at its terminal end.

Drumlins are asymmetrical, egg-shaped hills composed of till. Till is unstratified material of various sizes consisting of a mixture of clay, silt, sand, gravel, and boulders. The steeper side of the hill is the uphill side and can range from 50 to 165 feet (15–50 m) in height. The downward slope is much gentler and can extend up to 0.6 miles (1 km) in length. Drumlins usually form in groups, making them a distinctive depositional landform.

Eskers are long snakelike ridges formed by networks of streams that run under glaciers. As the rivers flow, they pick up and carry material, such as rock fragments. When the matter falls out of the water and is deposited along the channel, it forms an esker. Once the glaciers have melted, these features look like worms traveling across the landscape. Eskers can be 330 feet (100 m) tall and travel for 65 miles (100 km).

Larger pieces of rock deposited randomly on the ground's surface are called erratics. Erratics can range in size from pebbles to huge chunks of rock. They can be moved large distances when carried by the ice. Deposits of erratics help climatologists determine where ancient glaciers traveled.

Before an area is glaciated, mountain valleys are typically V-shaped because they have been carved by the erosive power of rivers. During glaciation, however, these narrow valleys deepen and widen under the erosive power and weight of the ice. The moving ice slowly grinds away at the canyon walls, weathering it smooth and creating distinctive U-shaped valleys, another strong indicator of climate change.

When a glacier first forms high up on a mountain, the initial bowl where the snow collects and begins the glacial cycle is called a cirque. A

Glaciers flowing out of canyons carve out a characteristic U-shape, as shown in this photo. Note also at the lower left corner, the large granite boulders strewn along the ground. They were carved by the glacier farther up the canyon, carried downhill, and deposited at the mouth of the canyon as glacial erratics.

cirque glacier is a small glacier that occupies a cirque or rests against its headwall. If a mountain peak has more than one cirque and is eroded from more than one cirque at a time from different directions (such as back to back) at its peak, it can form an arête. An arête is a narrow crest with a sharp, knifelike edge. If three or more arêtes converge, it forms a pyramid-style peak called a horn. One of the more famous of these glacial features is the Matterhorn in Switzerland.

Kame and kettle topography is another easily identifiable glacial landform. Kettles are depressions in the ground. They occur where large blocks of ice are caught in the glacial deposits. After the ice melts, large holes are left in the sediment. Kettles can vary in size, but most average about 1.2 miles (2 km) in diameter and 33 to 165 feet (10–50 m) deep. Kames are upraised deposit features that look like mounds or columns. Kames form when meltwater deposits sediments through openings inside the glacial ice.

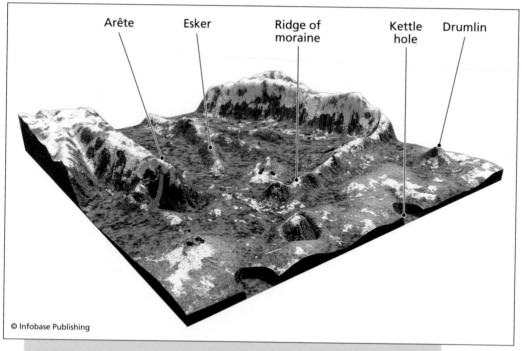

Arête   Esker   Ridge of   Kettle   Drumlin
                  moraine   hole

© Infobase Publishing

Glaciers are responsible for many distinct landforms, such as eskers, drumlins, moraines, kettles, and arêtes.

Another signal that glaciers have passed over an area is glacial striations. These are etched parallel grooves in rocks cut when a glacier drags material (such as rocks) over the rock it is sliding across, scratching the base rock as it passes over. These parallel etchings, or furrows, in the rock not only clue climatologists in to the fact that glaciers once existed in a particular area but also indicate which direction the glacier was traveling.

When these types of cold environment landforms are present in a landscape, climatologists look at past glaciation and reconstruct past climates. This is valuable information that can be used in climate modeling on computers.

Geological and Geochemical Proxy Data

The geologic record truly serves as a record set in stone of valuable information concerning climate change and the relationships between the Earth's physical systems and the life that occupies the planet. As climates change, they exact forces on the landscape that change the arrangement of the Earth's surface features, as illustrated in the previous chapter; but they also alter the Earth's surface and subsurface features in chemical ways. Although these alterations may not always be obvious, they also tell a detailed story about the climates of the ancient past. Because of this, it is possible to determine seasonal variations in temperature and precipitation, changes in the movement of ocean currents, and fluctuations in the carbon cycle going back hundreds of millions of years by analyzing geologic and geochemical proxy data.

Feasible proxies include the chemical and isotopic compositions of minerals as well as their physical properties. Proxies also include the gases and fluids contained in the oceans, in glacial ice, and on land.

Specifically, the ocean provides information about currents, salinity, nutrients, temperature, and chemistry. The land holds clues to temperature, precipitation, evaporation, and humidity as well as the forces of wind and weathering.

As technology improves, paleoclimatologists are able to uncover more data and analyze them more precisely. In the past few years, their understanding of ancient conditions has grown tremendously through the analysis of proxies. A combination of (1) more detailed information, (2) more refined age-dating techniques, and (3) the increasing availability of a worldwide distribution of data has allowed scientists to reassess old data, make new discoveries, and gain a much clearer understanding of past climate change, what caused it, and what its impacts have been.

When scientists use proxy data, they use statistical analyses to correlate the data and trends that show up in the proxies to the climatic characteristics that are directly observable and measurable today. Some analyses are made based on directly measurable physical characteristics; other conclusions are drawn on the similarities of relationships seen between evidence from the past and directly observable phenomena today. For example, if a paleorecord indicates that palm trees once grew in a polar area, it is assumed the area once had a much warmer climate because palm trees grow only in warm climates today.

Climatologists do know, however, that proxies serve merely as representations of past climates, so it is desirable to have more than one type of proxy representing the same climate interval in order to cross-check and ensure as much accuracy as possible. For instance, if there is both ice core data and fossil data from the same time period and they both confirm the same type of information, this adds to the validity of the climate reconstruction.

One of the most important issues in understanding past climate—just as it is today in understanding global warming—is how the climate reacts in response to changes in the concentration of greenhouse gases, especially CO_2. Understanding the effects that involve CO_2 is also important for being able to project future global warming and climate change. Climatologists know that the concentration of CO_2 in all Earth's reservoirs—the atmosphere, soils, and oceans—directly influences not only the Earth's climate but the oceans' pH and productivity as well as weathering and erosion rates on the Earth's surface. All of these complex components work together as a system.

This chapter will look at the geologic and geochemical proxies scientists use to determine past climate. First, the chapter will look at the Earth's distribution of soil and how climate can be directly linked to the type of soil that exists in a specific area. Then it will examine the role of sediments—both on land and underwater—and what clues it holds about past climate conditions. Next, it will explore ice cores and their importance in putting the climate puzzle together as well as how loess deposits are used and cave formations interpreted.

SOILS AND MINERALOGY COMPOSITION

Soil is composed of minerals (rock, clay, silt, and sand), air, water, and organic (plant and animal) material. The chemistry of soils and rocks is also affected by climate, and through the study of paleosoils (ancient soils), scientists can infer what past climate was like. In humid climates such as jungles, for instance, the soluble minerals are dissolved and washed out of the soil. Because only certain minerals do this, soils of humid climates are characteristically missing or deficient on these certain elements: potassium, calcium, sodium, magnesium, and silicon. Humid climate soils generally have characteristically large percentages of iron and aluminum because they are the least soluble and are the ones that tend to remain in place. Iron and aluminum form large deposits called oxides. Iron gives soil a red color, which is why a lot of soils in the Tropics have a rich reddish color.

In dry climates, water picks up small amounts of soluble minerals from the soil, carries them to low-lying flat areas, and deposits them. Then, under the hot desert sun and extremely low humidity, the water evaporates, leaving the soluble minerals behind. Over time, these minerals accumulate and form deposits called evaporites. Desert evaporites are generally composed of a mixture of sand, gypsum, various salts (sodium chloride and sodium nitrate), and borates. Evaporite mineral deposits are often found in playas and closed basins. The Great Basin desert of the United States is a major source of evaporite resources. These resources are considered mineral deposits and are mined commercially. Death Valley in California and the Bonneville Salt Flats in Utah are both well-known examples of evaporite deposits. Boron, obtained from borate evaporates, is mined from Death Valley and is a key ingredient in the manufacture of ceramics, glass, agricultural chemicals, water softeners, enamel, and pharmaceuticals. Salt and magnesium chloride are mined from the Bonneville Salt Flats area around the Great Salt Lake in Utah. The Bonneville Salt Flats are also home to the world-class car racing speedway.

Many types of soils exist on Earth, and each has its own unique properties, such as texture, color, structure, and mineral content. The depth of the soil also varies, partly determined by climate.

The soil-forming process is extremely slow. The parent material (rock) must first erode into small pieces in and on the ground. During this lengthy process, organic matter decays and mixes with the rock to slowly form soil. Developing soil is made of distinct, identifiable horizontal layers called horizons, which vary in characteristics. The deepest horizons (subsoil, regolith, bedrock) most closely resemble the parent materials, while the surface layers (humus, topsoil) contain the most organic nutrients, which is why they are fertile enough to support plant life.

Because there are several different soils that exist worldwide and climate has a large influence on their formation, pedologists (soil scientists) have classified them into 12 soil orders. Worldwide, there are various types of classifications. In the United States, soils are classified based on the concept of diagnostic horizons.

The breakdown below shows the typical layers of a soil profile. They are usually divided into the O, A, E, B, C, and R horizons with the following characteristics.

O: This horizon is on the surface and is the richest in nutrients, which is where most plant life and interaction of life-forms exist. It contains decomposed organic matter.
A: This horizon is the topsoil. This is where seeds germinate and plant roots thrive. It is composed of humus and mineral particles.
E: This layer is lighter in color and is composed of mostly sand and silt, because most of the clay and minerals have been eluviated (leached) by water as it flows through the soil under the force of gravity.
B: This layer is the subsoil and contains clay and mineral deposits, such as iron, aluminum oxides, and calcium carbonate that wash down into it from the layers above.

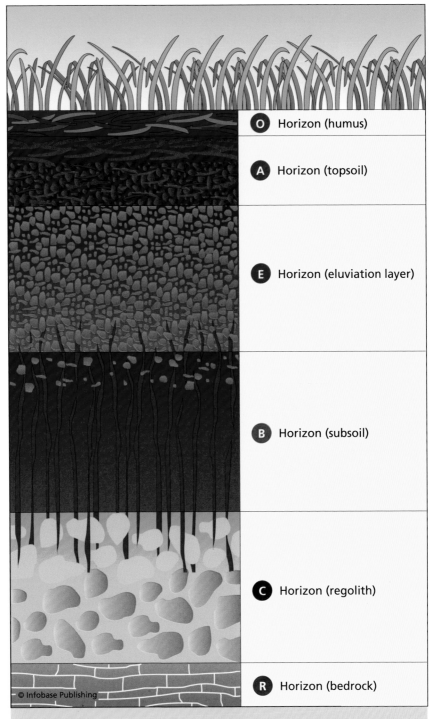

O   Horizon (humus)

A   Horizon (topsoil)

E   Horizon (eluviation layer)

B   Horizon (subsoil)

C   Horizon (regolith)

R   Horizon (bedrock)

© Infobase Publishing

A soil profile consists of distinct horizons. The climate plays an important role in the type of soil that exists in an area.

This soil profile in central Iowa shows the depth of topsoil—the rich, dark color is indicative of fertile soil. *(Lynn Betts, Natural Resources Conservation Service)*

C: This is the regolith and is made up of broken rock fragments. Organic matter is rarely deposited in this horizon.

R: This is the unweathered bedrock that the soil profile sits on.

Soils serve as a valuable diagnostic tool for climatologists, because different soil types develop in different climates. In forest biomes, the soils have a light gray upper horizon, a horizon rich in aluminum and iron, and form in humid regions that generally range from warm to cool, allowing conifers to grow. In grassland biomes, such as prairies, these rich soils have a dark surface layer and are typically rich in minerals. Today, these soils are common in the Earth's midlatitude regions. When ancient versions of these soils are found, it indicates what the climate was like when the soils formed.

In lowland regions, such as swamps and wetlands, the soils have a high water content and are poorly drained. The soils in these zones are dark-colored organic soils typically rich in organic matter. As illustrated earlier, desert soils are usually rich in calcium carbonate. In tropical zones, where it is humid and warm all the time, soils are a reddish color and extremely rich in iron oxide. They are depleted in nutrients. In fact, in the world's rain forests, most of the nutrients are contained within the forest vegetation, not the soils. In cold environments, such as the Arctic tundra, soils have a dark organic-rich upper layer and a mineral-rich layer that covers frozen ground.

There are 12 soil orders. By identifying the soil—or paleosoil—much can be deduced about the climate that once existed. The following table lists the 12 soil orders and their characteristics.

SEDIMENTS

Sediments, pieces of worn rock and other natural debris, both on land and from the bottom of the oceans, tell scientists a story about what past climate conditions were like on Earth. On land, sediments can become trapped in snow and ice; in the oceans, they are trapped deep on the floor of the ocean. Sediments at the Earth's surface can be preserved over time by being deposited, undisturbed and cemented together over the years, and then subjected to great heat and pressure, when they are transformed into layers of rock.

Changes in the ways that air masses interact with one another over decades, centuries, millennia, or even longer can vary so significantly that the effects can become evident in marks left on sediments and rocks. According to a study conducted by members of the U.S. Geological

SOIL	DESCRIPTION	PREFERABLE CLIMATE
Alfisol	Moderately leached soil with a subsurface zone of clay accumulation and $>= 35\%$ base saturation.	Temperate and tropical regions
Andisol	Soil formed in volcanic ash	All temperature regimes
Aridisol	$CaCO_3$-containing soils of arid environments	Both hot and cold arid and semi-arid climates
Entisol	Soils with little or no morphological development	A variety of climates
Gelisol	Soils with permafrost within 6.5 feet (2m) of the surface	Alpine and Polar Regions where temperatures are continuously at or below freezing
Histosol	Organic soils	Various climates
Inceptisol	Soils with weakly developed subsurface horizons	A variety of climates except arid ones
Mollisol	Grassland soils with a high base status	Various climates
Oxisol	Intensely weathered soils of tropical and subtropical environments	Zones with small seasonal variation and no soil freezing
Spodosol	Acidic forest soils	Humid areas
Ultisol	Strongly leached soils with subsurface zone of clay and $< 35\%$ base saturation	Where precipitation exceeds evapotranspiration and water storage capacity; found in tropical areas
Vertisol	Clayey soils with high shrink and swell capacity	Most major climate zones

Survey (USGS), sediments in the lakes in the upper Mississippi basin of the United States can provide a wealth of information on climate change for that area. They have documented both the Holocene paleoclimatic and paleoenvironmental changes. In their study, they looked at the various deposited sediments, the chemical characteristics of the lake beds, and the biological components of the area in order to reconstruct past conditions. They found that the best records came from lakes because their sediments generally contained identifiable annual increments of sediment.

One lake they visited and analyzed was Elk Lake in the Mississippi headwaters region of northwestern Minnesota, which had a nice sequence of varves associated with it. A varve is an annual layer of sediment. The Elk Lake varves they retrieved held an extraordinary amount of data about biological, chemical, and mineralogical components that they could relate not only to past climate but to environmental changes as well. Most of the sediments were lacustrine (formed in the bottom of the lake), but small amounts of silt and clay were blown into the lake by the wind and were mostly composed of quartz, making them identifiable from the lacustrine sediments. Both types of sediment layers were measurements of windiness. Diatoms are organisms that have heavy silica shells and require a certain level of turbulence to keep them in the photic zone (which also depends on windy conditions). Deposits of these were also analyzed in the sediments, indicating windy conditions. In addition, most of the sodium in sediments found in the deposits was blown in from soils surrounding the lake. Pollen was also blown into the lake.

Using all these proxy data, scientists were able to conclude that during the Holocene, after the last ice sheet retreated into Canada about 10,000 years ago, the area was first covered by a spruce forest and then replaced by a pine forest as soon as the climate began to warm. Then, about 8,500 years ago, climate continued to change and became drier. The pine forests were replaced by prairie vegetation (grass, sagebrush, and oak). During the next 4,500 years, the prairie ecosystem spread eastward 93 miles (150 km). Proxy information from soil samples taken in the area also confirms this. The sediments contained higher concentrations of sodium, indicating dry, drought conditions. Based on the proxy data collected and interpreted, experts at the USGS determined that this mid-Holocene drought affected a substantial area of North America and that active sand dune fields existed in Minnesota, the Dakotas, and Nebraska. One possible explanation for this is that a dry Pacific air mass controlled the climate of North America for a long period, a condition very similar to that experienced in the United States during the dust bowl of the 1930s, but on a much larger scale.

Worldwide, the remains of plants and animals, particles of matter, rock, dust, and clay accumulate at the bottoms of water bodies, such as oceans and lakes. Sediment deposits gradually build up over the years. In order to study the sediments, scientists must be able to retrieve them intact from their underwater environments. To preserve the soil profile, it must be removed intact. To accomplish this, a cylinder—also called a core—is extracted. The sediment core is then taken to a laboratory, and the contents are analyzed for the proxy evidence contained within them.

When ice cores are collected from deep ocean environments, the sediment cores within them may provide much useful information about past climate changes. Sediments can be infiltrated with foraminifera (also called forams), extremely small marine animals. They are useful because they are good proxies for past ocean temperatures. When forams die, their shells sink to the ocean floor and become buried in the sediment. Because different types of forams live in different types of water temperatures, the species found in sediment cores can be used as proxies to infer the climate and temperature during the time they lived. In addition, when alive, forams take up two isotopes (two different kinds) of oxygen in order to grow their shells. The type used is determined by the temperature of the water. Because of this, the oxygen isotopes also act as proxies. By measuring the relative concentrations of the oxygen isotopes in the shells, scientists can verify and estimate past water temperatures, and hence climate conditions.

Cores are also retrieved from lake bottoms, a specialized field of science called limnology. All day every day, sediments accumulate. Embedded in these sediments are records of organisms that lived in and around the lake as well as proxy data that clue in scientists to the processes that occurred within the lake, the composition of the lake's water, the conditions of the watershed, and the atmospheric conditions that existed at the time of deposition.

The most commonly used proxies in paleolimnology are diatoms, which exist worldwide in almost every body of water, although each type is unique and identifiable. Some species prefer acidic waters and others alkaline waters. Some are found in nutrient-rich or nutrient-depleted environments. These differences help paleoclimatologists reconstruct past climate conditions and are desirable proxies because they are sensitive indicators of environmental conditions.

Plant remains are often found in the cores taken from lakes. These serve as extremely beneficial proxies because certain plants grow only in certain temperature ranges, precipitation ranges, and biomes. It is not uncommon to find leaves, wood, seeds, twigs, and pollen in the sediments at the bottom of a lake. Once the different types of vegetation are identified, scientists can determine what the climate was like in the area when the vegetation was alive.

If a core contains a high amount of silt and clay, this informs climatologists the climate was very cold at one time, with sparse vegetative cover and increased soil erosion. As water washed across the exposed soil, it carried it into lakes, where it eventually settled to the bottom. Conversely, if sediment cores have high organic content, it signifies a warm and humid climate that once supported ample vegetation.

Experts worldwide analyze sediment cores in order to understand the Earth's ancient climate. For example, Will Sager, at Texas A&M University, studies ocean sediments. He works with the Ocean Drilling Program (ODP), which uses a ship fitted with an oil well drilling rig to retrieve sediment cores from the oceans' bottoms. One of his major research objectives is to study variations in rock magnetic properties that can be used as proxies to determine climatic variations. The ship itself is almost as long as two football fields, with a derrick that towers 200 feet (61 m) above the waterline. A long pipe fitted to the rig can descend through 4 miles (6.4 km) of water to drill holes into the ocean floor and pull out ocean cores that are 36 feet (11 m) long. A seven-story laboratory hosts facilities for the examination of sediment and rock cores, studied by scientists from 21 countries—physicists, biologists, chemists, geologists, geographers, climatologists, and others. The core must be brought to the surface very carefully; if disturbed, millions of years of data could be destroyed. Many of the samples are analyzed on the ship in the on-board laboratory. Researchers analyze the particles and chemicals that have been trapped in the sediment for millions of years, specifically minerals such as silicates and dead microorganisms, plankton, and diatoms. As the sediments were deposited and buried, water was slowly squeezed out, and the sediments eventually turned to rock.

According to the USGS, geological oceanographers study thin slices of the retrieved cores. If adequate amounts of proxy data are available—minerals and microfossils—they can infer past temperatures of the ocean and how similar or dissimilar it was to today's ocean conditions. The ocean provides a unique environment for this because, unlike on land, erosion is not an issue, and a continual settlement of material is supplied to the ocean floor, providing sediment that shows continuous records of the past.

Oregon State University is also involved in analysis of deep-sea sediments. It maintain a repository of ocean core samples that scientists access in their quest to better understand past climate change. For example, Alan Mix, a professor at the College of Oceanic and Atmospheric Sciences (COAS), has spent considerable time analyzing fossil shells from the ocean floor and their carbon and oxygen isotopes in order to better understand upper ocean temperature change through time—an important aspect of global warming. Mix also analyzes forams and the seasonal cycle of how they settle to the ocean floor. The ways in which they accumulate in the sediments can tell scientists how climate has changed over time. According to Mix, "Many foraminifera—forams for short—live near the surface of the ocean. We can get a measure of near-surface temperatures, which says something about upwelling systems that bring cold water up from underneath, from deeper in the ocean. Other foraminifera live on the seafloor, and they allow us to track the organic matter that falls to the abyss."

Mix's main focus is on Earth's past ice ages because they provide useful information about global-scale climate change. He believes that it is the large-scale, long-lasting changes that involve the entire ocean that humans can expect to face with global warming over the next hundred years. Mix refers to the ice ages as "a natural laboratory of sorts to understand processes of climate change. You need to understand both temperature and salinity, and we think we have steadily improving tools to do that. It will usher in a new age of paleo-oceanography when we can really deal with both those variables."

According to Mix, "You can't just go randomly out there and plop a core down and expect to get something good and useful. Analytical tools include surveys using sound waves to make maps of the seafloor and what is beneath it. Although most coring is successful, sometimes researchers have to go back over the years to get it right."

Another interesting observation Mix has made from his research on the ocean is that, while all along scientists have believed the ocean was a passive component of climate change—that it simply reacted to changes in the atmosphere as a passive absorber of climate change—this is not true. Instead, the ocean is a dynamic player in ice age–scale climate changes and is much more sensitive to change than scientists originally thought.

Weathering rates from continents, deep ocean flow mapping, and oceanic thermohaline circulation changes are three areas being looked at today through the analysis of carbon-13 records. Scientists use the carbon-13 proxy to study the global carbon cycle. So far, carbon-13 research looks promising as a new proxy that will be able to shed light on past global warming.

ICE CORES

Another significant area by which climate proxy data are gathered is from ice cores. The most popular places for obtaining ice cores are in Greenland and Antarctica because they represent long histories of ice accumulation. Many ice cores are collected and then sent to and stored at the National Ice Core Laboratory in Denver, Colorado, where they are further analyzed. Ice cores contain a wealth of information about the climate. Ice cores can contain an uninterrupted, detailed climate record extending back hundreds of thousands of years. Valuable information can be obtained from them, such as temperature, precipitation, chemistry and gas composition of the lower atmosphere, volcanic eruptions, solar variability, sea-surface productivity, and other climate indicators. It is the simultaneity of these properties recorded in the ice that makes ice cores such a powerful tool in paleoclimate research. In order to obtain the ice cores, a sharpened pipe rotating on a long cable is drilled into the ice. Once the core is drilled, it is extracted and brought to the surface.

Each year, new layers of snow collect on the ice sheets, and each layer is different in chemistry and texture. Summer snow is different from winter snow. For instance, in the summer, it is daylight 24 hours a day in the polar regions. Because of this, the top layer of snow has a different texture (it does not melt, but it differs in texture from the layers beneath it). When it turns cold and dark again, an increase in snowfall occurs. This layer looks much different from the layer that formed during the summer months. Similar to the case with sediment from ocean and lake bottoms, ice cores take on a characteristic banded appearance that differentiates the seasonal accumulations into distinct layers. Over hundreds of thousands of years, this snow, then ice, collects and traps a record of past climate within it. As scientists retrieve these cores—some from depths of more than 11,000 feet (3,500 m)—they begin unraveling the ancient climate history of an area.

Sometimes scientists dig snow pits in the field, wherein the snow layers are even easier to distinguish. By digging two pits adjacent to each other separated by only a thin wall of ice, it is possible to catch the light shining through the ice and illuminate the individual layers deposited within the ice.

Ice cores have been drilled from ice sheets since the 1960s. Some of the ice cores have contained data as far back as 750,000 years ago. These cores provide an annual record of temperature, precipitation, what the atmosphere was like at the time, any volcanic activity that may have been nearby, and what the prevailing winds were. The thickness of each layer supplies information as to how much snow accumulated. When cores are taken from the same area but show differences where the snow drifted from one place to another, it gives scientists a good idea of what the wind and atmospheric circulation patterns were like.

When scientists analyze the ice in the cores themselves, they look at the oxygen isotopes because that is where clues to past temperatures

These are the annual accumulation layers of the Quelccaya ice cap, each accumulation layer about 2.5 feet (0.75 m) thick. The light layers are the snow accumulation layers in the winter. The dry season is differentiated with the darker dust bands. *(Lonnie Thompson, Byrd Polar Research Center, Ohio State University, NOAA Paleoclimatology Program)*

are hidden. Specifically, the ratio of oxygen isotopes relates to how cold the air was where the snow was deposited. Colder temperatures have higher concentrations of light oxygen.

Oxygen is a key factor in understanding past climate. Oxygen exists as a heavy and light isotope. Oxygen-16 is considered the light isotope, and oxygen-18 is the heavy one. Oxygen-16 is the more common of the two. The ratio of the two isotopes of oxygen in water changes with the climate. In determining the ratio of heavy and light oxygen in ice cores and marine sediments of fossils, scientists learn how the climate varied over time. Because the heavier oxygen falls out, this leaves the light oxygen frozen in the ice caps, signifying cooler temperatures during this time period.

Thickness of an ice sheet is another important factor because it makes the ice sheet's temperature more resistant to change. Scientists measure the temperature of an ice sheet by lowering a thermometer into a borehole drilled to retrieve an ice core. The ice cap has an insulating quality that keeps temperature of the snow and ice preserved at the general temperature of the atmosphere when the layer was initially deposited. Near the Earth's bedrock surface, however, the lowest layers are warmed by heat from the Earth. Knowing the Earth's temperature, scientists use this information to calibrate the temperature record they retrieve from the oxygen isotopes. Thus, they can accurately detect temperature variations that have occurred since the ice age.

Some of the most valuable information obtainable from ice cores is data about the atmosphere. When snow initially forms, it crystallizes around tiny particles of smoke, dust, volcanic ash, pollen, and other matter suspended in the air. These particulates become trapped in the snow and can provide clues about the environment from which they came. When snow settles on the ground, air fills the spaces between the ice crystals. Then, as new layers of snow fall, the older layers are buried. Layer upon layer accumulates, and the space between the crystals is sealed off, trapping small pockets of the atmosphere in the ice. These small bubbles remain intact, and when an ice core is removed, the air bubbles are analyzed to see what the atmospheric conditions, and hence the climate, were like at the time of deposition.

Scientists must extrude (extract) the ice core from its barrel with extreme care. The cloudy layers visible in this core section were formed when dust fell onto the ice sheet and was trapped in the ice. *(Mark Twickler, University of New Hampshire, NOAA Paleoclimatology Program)*

According to scientists at NASA, records of methane in samples indicate that a climate conducive to wetlands once covered the Earth. According to Gavin Schmidt at NASA GISS (Goddard Institute for Space Studies), methane has oscillated in response to rapid climate changes, such as the Younger Dryas cold interval and rapidly increases in a warming climate, with a small lag behind temperature. This means that not only does methane affect climate through greenhouse effects, it can also be directly affected by climate. Natural methane emissions depend on the extent of organic decomposition in very wet conditions. For an individual wetland, an increase in the water table and/or an increase in temperature will lead to greater emissions on a very short time scale. Over longer periods of time, wetlands come and go as a function of rainfall patterns. While methane was not considered a very important greenhouse gas in the past, its role is better understood today, and it is now acknowledged to be an important greenhouse gas thanks to emerging mathematical models. Wetlands support a wide variety of life-forms that encourage the growth of anaerobic bacteria, which release methane as they decompose. Analyzing ice cores also allows scientists to correlate the amount of CO_2 in the atmosphere with climate change. Being able to track the levels of CO_2 throughout Earth's history helps scientists better understand the issues surrounding global warming today.

In fact, when levels of methane and CO_2 found in ice cores are compared with those found today, scientists have determined that the levels are higher now than they have been in the past 220,000 years. Activities such as industrialization, agricultural practices, and deforestation are currently blamed for this situation. Ice core analysis also reveals that during the past 220,000 years, when the atmosphere contained low levels of CO_2 and methane, climate was cool. Conversely, when these levels were high, climate was warmer. Scientists have determined there is a strong correlation between the Earth's temperature and greenhouse gas levels.

When wind-blown dust is analyzed in ice cores, it can be chemically analyzed to determine where it came from. The amount and provenance of the dust provide important information for unraveling the past because it sheds light on wind patterns and strength. Volcanic ash is another particulate that can provide climate change information because volcanic eruptions themselves sometimes contribute to climate change. Significant eruptions can put enormous amounts of ash into the atmosphere, cooling the climate for a period of time. If the layers of volcanic ash can be dated in the ice cores, then a correct geologic time line can be constructed from the layers.

LOESS

Loess is a homogeneous, fine yellow soil that has been deposited across 1 million square miles (2.6 million sq. km) of land that covers several areas of the world: Asia, Europe, and North America. It ranges in thickness from area to area and can be as thick as 10 feet (3 m) in some locations. Loess (rhymes with *us*) originated from glacial processes. As the massive weight of the glacial ice moved across the Earth's surface, the ice ground along the rock slowly and abraded and pulverized it into a powderlike substance. Later, as the climate warmed and the ice melted, running water washed the flourlike deposits from under the glaciers and into streams along the edges of the ice. As the area dried out, winds carried loess across the land. Its texture was so fine that it was carried great distances. This spread the deposits across wide areas and left rich, easily recognizable, homogeneous soil.

Loess is a proxy because climatologists can use it to map the past existence of glaciers. In the United States, for example, scientists have been able to determine through the mapping of both loess and rock debris deposits that ice sheets once existed in the area where the Great

This is the upper part of a section of loess near Baoji in southern China's Central Loess Plateau. Climatologists have determined that these soils represent interglacial periods when climates were wet enough to sustain vegetation development. During glacial periods, climates were colder, drier, and windier, leading to sparse vegetation cover and extensive mineral dust (loess) accumulation. *(Nat Rutter, Department of Geology, University of Alberta, Edmonton, Canada, NOAA Paleoclimatology Program)*

Lakes are located today. Ice sheets also covered the British Isles and Scandinavia in Europe during this period, as evidenced by loess deposits there.

Loess deposits have also been found in mountain (alpine) environments, indicating the existence of glaciers in the past. Evidence has also been uncovered of multiple glaciations using loess as a proxy tool, as some loess deposits are much older. Currently, climatologists believe there have been at least three or four individual ice events that generated loess.

CAVE ENVIRONMENTS

Unique rock formations in caves display a climate record of their own and serve as proxy indicators of past moisture conditions. In the southwestern United States, more than 100 caves exist whose formations tell a story of what the climate was like long ago. In addition, because the formations in caves are preserved underground, they are protected from the harsh weathering and erosional processes to which features on the Earth's surface are subjected.

The wonderland of geological formations found in caves is formed when water soaks through the ground and picks up minerals, the most common being calcium carbonate. When the water drips into caves, it is laden with calcium carbonate. As the water drips, it leaves behind the mineral deposits, which are the same types of hard white deposits that collect on sink faucets and showerheads in homes. As the minerals accumulate, they form the iciclelike rocks that hang from cave ceilings called stalactites. Minerals drip off the stalactites and are deposited on the ground directly below them, creating another deposit from the ground up called a stalagmite. The calcium carbonate can collect in many interesting shapes, such as flowstone and popcorn.

Caves are another proxy of past climate change because they are indicators of past water conditions, which provide important information about past rainfall and temperature. They are formed by the subsurface action of water, like a type of huge subterranean plumbing system. Many massive subterranean cave systems, such as Carlsbad Caverns in New Mexico, exist throughout the world, through which water does not currently flow but did in the past.

Cave formations give scientists clues as to what past climatic processes were like in an area. This is the Chinese Theater formation located in Carlsbad Caverns National Park, New Mexico. *(National Park Service)*

Caves initially form from the dissolving of carbonate rocks and the formation of cavities and passageways. This occurs in an area just below the water table in the zone of saturation, where the continuous mass movement of water occurs.

The second stage in cave development occurs after the water table lowers. During this stage, the solution cavities (the hollow openings left after the removal of the calcium carbonate in limestone by carbonic acid) are abandoned in the unsaturated zone, where air can enter. This leads to the deposition of calcite, which is what forms the dripstone features.

The mineral formations in caves are collectively referred to as speleothems. As long as water is actively flowing into caves, speleothems continue to grow in thin, shiny layers because mineral deposits are continually being added. The amount of growth they experience is a direct indicator of how much groundwater percolates into the cave. If there is only a small amount of growth, this may indicate that the existing climate was dry. Conversely, rapid growth may indicate that the climate at the time of deposition was extremely wet.

Geologists can also tell the difference between when speleothems are actively growing and when they stopped growing. When they are active, they have a smooth, wet appearance. When they have stopped growing and are inactive, the outside becomes dirty and eroded, making it look dull.

As speleothems form and grow layer upon layer, scientists have been able to date the individual layers in some caves by measuring how much uranium (a radioactive element used for dating; see chapter 3) has decayed. This is possible because uranium from surrounding bedrock seeps into the water and forms a carbonate that is incorporated into each layer as it forms. As the uranium (parent element) decays into thorium (daughter element), it adheres to the clay in the bedrock instead of escaping into the groundwater and into the speleothems. Because the thorium does not escape, the newest layer added to the speleothem contains no thorium. The uranium in the deposit then decays into thorium, enabling scientists to radiometrically date the deposit by measuring the ratio of uranium to thorium.

By dating each individual layer of a speleothem, paleoclimatologists can compile a record of how groundwater levels have changed over the lifetime of the formation. According to paleoclimatology experts at the National Oceanic and Atmospheric Administration (NOAA), however, using information from just one cave is not sufficient to infer climate conditions over broad areas. Scientists must look at caves in several geographic locations and search for similar patterns of growth in speleothems to accurately infer regional climate change.

One area scientists are looking into with speleothem research is analyzing oxygen isotope records (oxygen-18 and oxygen-16). Through the analysis of tiny bubbles in the speleothems that contain trapped air and water, scientists can look at the oxygen isotopic ratio to calculate past air temperatures. As discussed earlier, the ratio of these two isotopes in water varies based on air temperature, the amount of precipitation, and the amount of water locked in ice caps and glaciers worldwide. Scientists believe that the ^{18}O-^{16}O ratio in cave deposits may be able someday to be directly correlated to understanding global climate conditions.

To support this, scientists have been able to use this oxygen isotope ratio to determine the origin of rainfall in certain geographic areas, that is, whether it was from coastal or inland sources. According to NOAA, by analyzing the oxygen isotope ratio, it is possible to track seasonal changes and rainfall patterns, and since caves exist worldwide, speleothems may have the potential to play a key role in understanding the land-based global climate record.

Biotic Proxies

Climate proxies allow multiple independent pieces of a complex puzzle to be put together to create a master picture of the Earth's past climates. Each piece, no matter how small, makes an important contribution in the effort not only to understand when and where climate change occurred but also how and why. Knowing the answers to these questions is important because it strengthens the community's understanding of the sensitivity of the Earth's climate, how it responds to global warming, and the effect human behavior is having on natural systems. Another major group of proxies that scientists use to study past climate change are the biotic proxies. These include items once alive on Earth, such as life-forms, trees, coral, and plants. This chapter first looks at how the presence of fossils and other evidence of once-living creatures tell a story about the climate of the past. By examining them and the associations of environmental conditions today in which their modern-day descendants live, climatologists can gain clues as to what the past environment was like. Next, it looks at trees and the science of dendro-chronology and examines how far back into the historical time line this technique can reliably take scientists when they attempt to reconstruct past climates. The use of corals is presented next, including the environment they need in order to survive and why ancient coral formations are found today in some of the most unexpected places. Finally, the paleoclimatic story told by plants and how they relate directly to climate change is examined.

LIFE-FORMS/FOSSILS

Various life-forms are used by paleoclimatologists to reconstruct the Earth's climate changes through time. They include fossil evidence, forams (foraminifera), pack rat middens, and insects. Fossils are important because they provide a conclusive record as to what animals and other life-forms existed in a specific geographic environment during a specific period. Fossils of forams, in particular, are especially helpful because they have broad geographical ranges. Pack rat middens (crystallized pack rat urine) clue climate scientists in to a specific type of climate and have played an important role in the reconstruction of the southwestern portion of the United States. The presence of insects also provides important clues to the past environment. Based on information available on insects and the types of climates in which they live today, it is possible to reconstruct past climates when remains of insects are found in the ancient record.

Fossils

One of the key ways scientists have been able to identify climate change is through the discovery of fossils. During the last ice age, for instance, animals indigenous to cold habitats lived much farther south than would be possible today, such as reindeer and wolverine. Likewise, scientists know that musk ox, a cold-climate species that currently lives in the Arctic, roamed as far south as Mexico City during the last ice age. A mastodon tooth was discovered off the coast of New Jersey. This type of proxy data tells scientists two things: (1) a species found in cold climates existed as far south as New Jersey, and (2) if the fossil was found offshore, it was not covered by ocean at the time the mastodon was there, which indicates that the ocean level was lower because much of the water was locked up in glacial ice and ice caps.

Fossil remains of wooly mammoths have been found in sand and gravel glacial deposits in North Dakota. *Bison latifrons* were recently discovered in North Dakota from the ice age, Pleistocene giant bison. Larger than today's bison, their horn cores spanned more than 7 feet (2 m), compared to the bison of today with a 2-foot (0.6 m) span. Paleontologists have also found walrus remains (an arctic animal) off the coast of Virginia, which means that walruses migrated that far south during the last ice age.

Forams

Some of the best fossils to use as proxy data are forams and diatoms. They have been used worldwide to piece together the picture of past climate. Forams and diatoms are shelled microorganisms found in aquatic and marine environments. One of the factors that makes them so versatile is that there are benthic types (those that are bottom dwellers) and planktonic types (those that float in the water), enabling them to represent a large range of the ocean environment.

The shells of forams are composed of calcium carbonate ($CaCO_3$). Diatom shells are composed of silicon dioxide (SiO_2). The shells of these microorganisms are important for recording past climates. Paleoclimatologists study the remains of foram and diatom shells deposited in sediment cores removed from lakes and oceans. (When the microorganisms die, their shells fall to the bottom sediment and are buried.)

Through analysis of the shells, the chemistry of the water can be derived. The oxygen isotope ratios (^{18}O-^{16}O) in their shells can be analyzed to determine what the water temperature was. As discussed earlier, warm water has less ^{16}O, which means that shells that lived in warm water are higher in ^{18}O. Through the collection of thousands of samples of forams and diatoms worldwide, scientists have been able to reconstruct past surface and bottom water temperatures.

In addition to water temperature, it is also possible to make inferences about the environment based on the foram and diatom record. For instance, the species of foram or diatom as well as their abundance

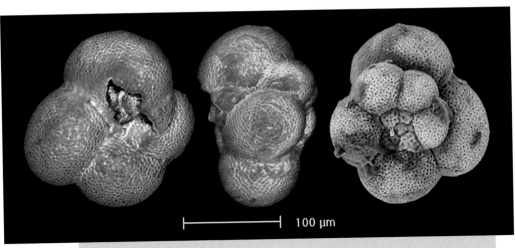

This foram is an *Eoglobigerina operta*. These fossils are commonly used for paleoclimatic dating purposes. *(University of California Museum of Paleontology)*

tells its own story. Scientists know that in warm climates the microorganisms multiply. Even more helpful, each species exists in its own set of desirable conditions, so when that species appears in the fossil record, it can be inferred that that particular condition existed at the time. Because tropical, subtropical, and polar species differ, their fossilized remains in deep-sea sediments allow marine geologists to map and date their relative distributions, thereby inferring variations in sea surface conditions caused by climate change and global warming.

Deposits of forams have also enabled paleoclimatologists to make assumptions about ocean circulation and major ocean currents. One specific foram, *Neogloboquadrina pachyderma,* is found only in polar regions. One form of the species is generally abundant in the coldest waters of the North Atlantic and lives only where the water is colder than 46°F (8°C). The other form exists only where the water is warmer than 46°F (8°C). This distinction enables scientists to track the 46°F (8°C) isotherm (the boundary marking the dividing line between where it is warmer or colder). This is important because it is directly related to the location where deepwater convection occurs in the North Atlantic's heat-carrying currents. Through mapping the foram fossil record, scientists have concluded that ocean circulation is indeed closely linked to rapid changes in climate.

Pack Rat Middens

Another proxy method that scientists use to reconstruct past periods of climate change is through analysis of plant remains from fossil pack rat middens. The U.S. Geological Survey (USGS) is currently involved in research along these lines in the southwestern United States. Fossil pack rat (also known as wood rat) middens (crystallized urine) often contain abundant fossilized remains of leaves, twigs, fruits, seeds, bones, shells, and other dateable materials. These findings help reconstruct the past environment, illustrating what the climate conditions were at the time. Information about past atmospheric conditions is also contained in the middens. Scientists derive this data through analysis of the ratios of stable isotopes of oxygen, carbon, and deuterium. According to the USGS, radiocarbon dating has identified some middens that are more than 50,000 years old.

Because these samples are useful so far back in time, their analysis has become important for reconstructing biotic communities and environmental conditions. To date, the USGS has been involved in pack rat midden and climate change studies in Arizona, California, Colorado, and Utah.

In a cave in Colorado, other researchers—Tony Barnosky, a paleobiologist from the University of California, Berkeley, and Chris J. Bell,

Pack rats are roughly the size of laboratory rats. They gather plant materials at close range and accumulate them in dry caves and crevices. The plant remains and other debris become cemented into large masses of crystallized urine that can survive for tens of thousands of years and be used as climate proxies. *(NOAA Paleoclimatology Paleo Slide Set; Ken Cole, USGS, photographer)*

Paul Martin of the University of Arizona is examining a pack rat midden. It was not until the 1960s that paleoecologists realized that middens could be used as a tool to reconstruct past climatic conditions. *(NOAA Paleoclimatology Paleo Slide Set; W.G. Spaulding, photographer)*

an associate professor of geology at the University of Texas at Austin—discovered a pack rat collection of teeth and bones that dates back to between 600,000 and 1 million years ago. Due to the arid cave's controlled temperature and humidity, the specimens were well preserved. According to Tony Barnosky, "Everything in the cave has been nicely preserved at a controlled temperature and humidity, like putting the stuff in a refrigerator for 750,000 years. This is the first study where we've actually taken a living species and looked back almost a million years at the population level to see how it changes through time. Our study suggests that species adapt to handle routine climate change, and only something out of the ordinary initiates significant evolutionary change. It takes a long time for a species to change, and even the major climatic change 800,000 years ago wasn't dramatic enough to cause the origin of a new species."

Based on his studies, Barnosky concludes, "It's likely that speciation takes place over a longer time interval than extinction. So, climate changes like the global warming we are seeing today are probably happening too fast to cause anything but extinction."

In their studies, they have analyzed environmental conditions of the area over a large span of time and have identified evidence of climate changes, such as glacial and interglacial events. A major climate change that took place 800,000 years ago was identified. One of the interesting conclusions they made is that of species evolution and adaptation. Through past glacial and interglacial episodes, they were able to detect adaptations in certain species, such as voles.

One thing they caution is that these natural adaptations took place due to the climate over long periods of time. Today, however, with global warming, climate change is happening much faster than it has in the past; some species may not be able to adapt quickly enough and may face extinction. Pack rat midden studies in North American deserts, especially in caves and rock shelters of the arid interior, are currently under investigation by several universities as valuable proxies by which to study climate change and global warming.

Fossilized insect remains are also used as paleoclimatic proxies. A study in British Columbia, Canada, conducted by S. Bruce Archibald and Rolf W. Mathewes in 2000 determined that in the early Eocene, several different species of insects existed: bees, ants, March flies, diplopterine cockroaches, dinidorid bugs, and seed weevils. Some of these insects were larger than their cousins today. Archibald and Mathewes concluded that because of their presence in the fossil record, the Eocene in British Columbia was much warmer than it is today.

According to the National Oceanic and Atmospheric Administration (NOAA), the beetle (classified in the order coleoptera) is commonly found in both freshwater and land sediments. There are several different species of the beetle, each preferring a specific type of climate. Because scientists know what climatic conditions the corresponding present-day beetle species like, as well as dating the sediments the beetles were buried in, scientists can infer what the climate was once like using this proxy evidence.

For more than 1,000 years, scientists in China have monitored the abundance of locust populations. They have long been aware that locust outbreaks are most common when the climate is cold and wet. They have linked locust records with temperature and precipitation from 957 to 1956. Their results match the climate record of both floods (high locust populations) and droughts (low locust populations) in the lower Yangtze River.

According to Marianne S. V. Douglas, Roberto Quinian, and John P. Smol, in the Canadian high Arctic sedimentary records in three ponds on Ellesmere Island show major shifts in pond communities over the past 200 years. The results indicate that aquatic insect populations greatly increased in size and diversity in the early 1800s. They believe these changes happened because of shifts in algal (diatom) populations, which indicate an enhanced food chain. The increase in food supply was most likely due to climate warming, which reduced the ice cover in ponds. They predict that future warming in the Arctic may cause even more drastic ecological changes. According to the authors, the multitude of shallow ponds that have existed in the region for millennia are now completely drying out during the polar summers. Through the comparison of measurements of pond water to those made in the 1980s, the disappearance of the ponds is attributed to increased evaporation-precipitation ratios. They believe this is due to global warming. They state that the "final ecological threshold for these aquatic ecosystems has now been crossed—that of complete desiccation."

DENDROCHRONOLOGY

Dendrochronology is the study of past climate change through examination of tree ring growth. Andrew Ellicott Douglass from the University of Arizona first used this specialized branch of science in the early 1900s. Douglass was the first to realize that the wide rings of certain species of trees were produced during years with ample rainfall and favorable growing conditions.

Every year, a tree adds a new layer of wood to its trunk and branches, which create the annual rings that are visible when viewing a cross section of a tree trunk. The new layer of wood grows from the cambium layer between the old wood and the bark. During new spring growth

Each year a tree adds a new ring, enabling scientists to determine the age of trees. Wet years result in thicker rings, dry years are associated with thin rings. By using these rings as a time line, climatologists can determine the past climate history of an area. *(Nature's Images)*

when rainfall is abundant, the tree uses its energy to produce new growth cells. The newest cells generated are the largest of the annual layer. As the summer progresses, temperature rises, and moisture become less abundant, the size of the cells decrease. Then, in the fall, the growth stops and the cells die. The tree remains dormant during the cold winter months. It is the contrast between the smaller old cells and the following spring's larger new cells that creates the visible boundary that is referred to as a ring. Each year, the tree repeats this cycle, adding a new ring. Over time, the tree becomes a measuring stick of time because the rings can be counted in order to determine chronology (time). Trees at middle and high latitudes produce the annual rings necessary to analyze past climate. Patterns in the width, wood density, and hydrogen and oxygen isotopic composition of tree rings can be used to estimate temperature.

Because the same set of environmental factors influence tree growth throughout a region, the patterns of ring characteristics, such as ring widths, are often common from tree to tree. These patterns can be matched between trees in a process called cross-dating, which is used to assign exact calendar year dates to each individual ring. The calibrated rings from a number of trees in a region are combined to form a tree-ring chronology. *(NOAA Paleoclimatology Paleo Slide Set, Laboratory of Tree-Ring Research, the University of Arizona)*

Climatologists study climate changes and climate patterns in geographic areas through analysis of tree rings. Samples from trees of unknown age can be studied to see if they match with trees that have been analyzed and dated. If cross-samples can be matched somewhere in their ring sequences and overlap in age with rings that have been dated, climatologists can look further into the past. This stair-step concept is referred to as extending the chronology using cross-dating techniques. Samples are usually taken from trees with a long slender boring tool without harming the tree. This technique has allowed scientists to establish some bristlecone pine chronologies in North America that

This is a bristlecone pine in the White Mountains of California. The bristlecone pines of the Great Basin region of the western United States are the oldest known living trees—reaching ages up to 7,000 years old, providing extensive historical records of climate. *(NOAA Paleo-climatology Paleo Slide Set; Jonathan Pilcher, Palaeoecology Centre, Queen's University, Belfast, photographer)*

date to nearly 9,000 years ago. Trees can grow to be hundreds to thousands of years old and can contain reliable records of climate for centuries to millennia.

It is important to keep in mind, however, that as with all proxy data, there are limitations to techniques. In dendrochronology, the most serious limitation is that more variables affect tree growth and health than just moisture availability. Tree growth itself is very complex, as is interaction with the surrounding environment. Growth of trees can also be affected by wind, soil properties (nutrients, physical structure), slope (steepness of the terrain), sunlight availability, temperature, winter snow accumulation, fire, and disease. All of these factors contribute in some way—big or small—to a tree's annual growth and therefore to the appearance of its rings.

When climatologists sample trees in the field, they do not base their conclusions on just one core from one tree. They usually take more than one core from each tree, and they sample many trees over a broad area.

When trees grow to become hundreds or even thousands of years old, they are a valuable source of information about climate change. When scientists can look hundreds of years into the natural record found in a tree's rings, they can unravel the clues necessary to answer questions today about the causes and effects of global warming. *(NOAA Paleo-climatology Paleo Slide Set; Peter Brown, Rocky Mountain Tree-Ring Research, Fort Collins, Colorado, photographer)*

Once they collect these data for an entire climatic region, they analyze them and take the average in order to obtain the best estimate of climatic conditions during a certain period. They then use computer programs to analyze the collected data.

The best trees to use for studying climate change are bristlecone pines, which can be thousands of years old, because as they age they are not as easily affected by environmental variables as are young trees. They are sturdier and can be successfully used to document the effects of global warming.

Statistical analysis allows climatologists to see overall changes in climate trends and address various environmental problems, such as global warming. Examining the sensitivity of trees to changes in climate allows not only past reconstruction, but also present-day monitoring, especially in high-latitude areas such as Alaska, Siberia, and Scandinavia, where global warming is already causing serious problems.

A. E. DOUGLASS (1867–1962): THE BIRTH OF DENDROCHRONOLOGY

Andrew Ellicott Douglass was an American astronomer whose major accomplishment did not involve space: He discovered dendrochronology. When he was just 27 years old, he found what he believed to be a connection between climate and plant growth. He first began recording the annual rings of pine and Douglas fir trees around his home. Then, in 1911, he discovered matching records among trees 50 miles (80 km) away. He invented the term *dendrochronology*, which means "tree-ring dating."

His discoveries made it possible for many other scientists to advance in their own fields—archaeologists could use it to date past civilizations, climatologists could use it to pinpoint significant climate events, biologists could date insect plagues, geographers could study past droughts, and geologists could interpret past volcanic events. His techniques allowed cross-matching and linked together "living tree" chronology with "archaeological tree" chronology.

As his work grew in volume and in popularity, a tree ring laboratory was built at the University of Arizona. Today, this laboratory contains the world's largest accumulation of tree ring specimens from both living trees and ancient timbers. Built under the football stadium at the University of Arizona, the Laboratory of Tree Ring Research (LTRR) preserves almost 1 million wood specimens used to study important scientific issues, such as climate change and global warming.

Through Douglass's efforts, scientists have been able to map climate history in the United States hundreds of years before European settlement. A true pioneer, Douglass also believed that dendrochronology was a valuable tool for measuring future climate.

CORAL

Coral reefs record small variations in the climate that can tell scientists about growing conditions in the oceans. Corals that grow near the ocean surface provide annual records of tropical climates that extend back over the past several centuries. They therefore serve as proxies of the upper ocean environment for sea-surface temperature and salinity.

When scientists want to inspect a core of coral, they must carefully extract a sample using a hydraulic drill connected to a compressor on a ship. Their goal is to drive a perpendicular core path into the coral so that they obtain a sample of the coral's maximum growth record. *(NOAA Paleoclimatology Paleo Slide Set; Maris Kazmers, SharkSong Photography, Okemos, Michigan, photographer)*

This is a positive X-radiograph collage of a Galápagos Pavona clavus coral. It is a sample of a coral skeleton composed of calcium carbonate ($CaCO_3$). A coral skeleton formed in the winter has a different density than one formed in the summer because of variations in growth rates due to temperature and cloud cover conditions, creating growth bands. The bands are made more visible in an X-ray. *(NOAA Paleoclimatology Paleo Slide Set; Jerry Wellington, department of biology, University of Houston, photographer)*

Corals build their hard skeletons from calcium carbonate they extract from seawater. The calcium carbonate contains isotopes of oxygen and trace metals scientists can use to determine the temperature of the water in which the corals grew. Similar to the way annual rings on a tree form, the living tissue of corals is generated on the outermost layer, which is where the annual growth band occurs. As with dendrochronology, it is the thickness and thinness of the rings that climatologists use to infer past climate conditions. The coral responds to variations in temperature and cloud cover, and the thickness of the bands is determined by the ocean's temperature and salinity.

When the water is warm, coral growth accelerates, and the growth layer is wide and porous. Conversely, when the water is cooler, the layers are denser. In coral banding, the lighter bands are those laid down in summer during intervals of fast growth, and the darker layers are those formed during winter, when growth is slow. There is a temperature threshold, however, and when the water temperature gets too warm, it can actually harm the coral by drastically slowing its growth or killing it in a process called coral bleaching. This is a problem now affecting the world's natural reef systems, and many scientists believe it is caused by global warming. Scientists also use chemical ratios within coral to determine past climate because ocean temperature has a direct correlation to coral's constituents.

According to scientists at NOAA, coral reefs today are in jeopardy globally, especially those located on shallow shelves near heavily populated areas. Not only are warmer temperatures damaging coral habitats, but so are human activities such as pollution, scuba diving, and boating.

VEGETATION

In geological formations, it is more common to find fossil remains of plants than of animals, meaning that vegetation plays a key role in the reconstruction of ancient climates. If a specific species of vegetation is found in a geologic formation that is tens of million of years old, it gives scientists an idea of what the climate was like. If a palmlike tree fossil is found at a high northern latitude, it can be inferred that conditions at one time were much warmer than they are today.

When climatologists attempt to reconstruct more recent climate records, they often look at pollen deposits in sediments. Each plant species produces its own uniquely shaped pollen grains. Tiny grains of pollen are produced in huge amounts and are then distributed by wind and deposited in lakes, where they become part of the permanent sediment record. Oxygen-free environments, such as peat bogs, are the best places to find pollen because they do not support decay. Different levels of pollen signal climate change. Scientists first identify the pollen by major type, such as tree, grass, or shrub, and then subdivide it into individual species, which usually carry important climate implications. Pine trees, for example, indicate a cool climate, oak trees a warm climate, and palm trees a tropical climate. Larger remains from vegetation can also be studied to infer climate type. Often, larger remains—what scientists call macrofossils—exist in an area, such as cones, seeds, and leaves.

Fieldwork in the northern Great Plains of the United States has revealed climate change through evidence of drought and fire proxy data in the pollen record. Some pinecones release their seeds only during fire incidents. It is known that Florida was affected by the last ice age based on fossil pollen deposits there. In northeastern Illinois, evidence has been discovered that dates back 17,000 years and confirms the retreat of the immense Laurentide ice sheet. Fossil pollen evidence exists that indicates the area was once covered with tundralike vegetation, a drastically different ecosystem than the prairie vegetation that existed when Europeans first came to America, not to mention the belts of corn and soybean that thrive there today.

Many plant fossils have been found in areas where it is too cold for them to grow today. Palms have been found, for instance, in Wyoming and Utah that lived 45 million years ago. Today they would freeze. Red horn coral, a very rare fossilized form deposited during the mid- to late-Cretaceous period 65 to 135 million years ago, has been found on what are today the high mountaintops of the Uinta Mountains in Utah. Living 65 to 85 million years ago, it is a unique coral that was deposited in an ocean. At that time, the Earth's volcanic activity forced new ridge systems to rise high above the old ocean depths in the Pacific Ocean and lifted neighboring ocean floors with them.

Using proxy data in this area, scientists have been able to determine that the climate was not only once warm enough to support the growth of coral, but that temperatures increased significantly when massive volcanic activity released enormous amounts of CO_2 into the air. The results were dramatic—the icecaps melted, and the oceans rose 656 feet (200 m) higher than they are today. The sea progressed inland through midwestern United States, almost into Canada, while much of Europe was under water. The sea covered a large portion of the Rocky Mountains, and because of the warming of the Earth's climate, it made an excellent habitat for coral to flourish.

According to the USGS, macrofossils (leaves, wood, cones, and seeds) have provided a wealth of information on how climate change affects Alaska's vegetation. Fossil records show that several significant changes have occurred in the vegetation at high (polar) latitudes during times of climate change. The fossils show that high latitudes are more sensitive to climate change than lower latitudes.

During the Miocene, a major global warming occurred about 17 to 14.5 million years ago. This warming drastically changed the vegetation composition. Instead of remaining forested with conifers, temperate species such as oak, hickory, beech, chestnut, and walnut moved into the area and dominated. Scientists at the USGS have calculated that during this period the temperature may have been 25–30°F (15–18°C) warmer than today.

Then the climate flip-flopped, and 14.5 million years ago, a global cooling effect began. The effect it had on Alaska's vegetation was not only abrupt but also dramatic. The temperate trees disappeared, and the conifers returned. Alaska experienced several of these heating-cooling cycles over the course of time, with similar results. USGS scientists believe temperatures during glaciations were 9–15°F (5.4–9°C) below today's temperatures.

By this switching off of cold-weather trees and mild-weather trees (and the shrubs associated with them), USGS scientists believe the fossil record is a valuable proxy that illustrates how past climate change has influenced Alaskan vegetation. They have discovered examples of climates both warmer and colder than the climate today, which has given them a better understanding of how various types of environments respond ecologically. Using this data allows them to test the results of new algorithms in computer modeling. Knowing that climate change and global warming affect species throughout an ecosystem, the USGS also sees its efforts at studying past climate as a way to predict the effects of future change on polar environments.

According to Thomas Ager at the USGS, "The study of past climates and ecological changes in Alaska are an important key to understanding the likely consequences of future climate changes in high latitude ecosystems. We can expect that future periods of cooler, drier climate will result in shrinkage of forest boundaries, lowering of altitudinal tree line, and expansion of tundra vegetation into lower elevations. A future change to warmer, moister climates will result in expansion of Alaska's forests into areas now occupied by tundra. The past record also shows that the magnitude of future global scale climate changes and ecological responses will be greater at high latitudes than at lower latitudes."

With all the proxy methods discussed in this and the previous two chapters, it is important to understand that there are limitations involved. Although technology has continued to progress and has made incredible advances over the past few years, there are still some major gaps in scientists' understanding of past and future climate behavior. Two very important components that affect weather and proxies that are not currently well understood are the properties of clouds and the composition of the atmosphere. As scientists continue to learn more about the complex system of climate, new windows, both to the past and future, will be opened.

Climate Change and Past Civilizations

Climate has played a prominent role in shaping civilization and cultures throughout time. Civilizations exist where they do because of several key features, many of which are directly related to climate. For instance, major cities are located where rainfall is plentiful enough to allow cultivation of crops and a supply of drinking water to large populations. They are located where temperatures are suitable for growing food and living comfortably.

Because climate can change, what may be suitable for the growth of a civilization at one time may not stay that way. If climatic changes occur, civilizations can be seriously affected to the point that inhabitants must adapt. If they cannot adapt, they face extinction.

Through the work of archaeologists, scientists know that many great civilizations have existed in the past, only to be obliterated when the climate changed and the civilization was not able to keep up. This chapter addresses some of these issues. It first looks at the role of climate on settlement patterns and why people choose to live where they do. It then focuses on the impacts of climate change on past great civilizations and their long-term effects. Finally, it looks at modern civilization and the adaptations required today in the face of climate change issues.

THE RISE OF CIVILIZATION

Increasing evidence points to recent global warming as a consequence of human activities, such as the burning of fossil fuels and deforestation coinciding with rising levels of CO_2 in the atmosphere. But scientists also know there are natural variations at work on the environment at the same time. One of the biggest questions climatologists must face is how to tell the difference between natural change and human impact as it occurs now and into the future. Experts point out that future change will require people to adapt and make educated decisions. They also suggest that in order to fully comprehend and predict future change, it is important to look at the experiences of past times and learn from them. By looking at the rise and fall of ancient civilizations, one may assess present vulnerability to future environmental change.

According to Gerald Haug, a professor of geology at the University of Potsdam, Germany, in an interview with National Geographic News, toward the end of their collapse, the Maya faced desperate times. Their cities were overcrowded, and agricultural production was not meeting their demands. Even though the Maya had successfully dealt with shorter droughts in the past, they were not prepared for the long-term drought that finally dealt their civilization its final blow. Haug said if the Maya had been able to hold out a couple of years longer, they may have survived; but they had no way to know the drought would eventually come to an end. Other societies have survived past climate changes by changing their behaviors in response to environmental change. Three centuries after the Mayan empire collapsed, the Chumash people on California's Channel Islands survived severe droughts by transforming themselves from hunter-gatherers into traders.

Haug says the Maya collapse can serve as a good lesson today. When droughts happen, they cause several hardships: crop failures, malnutrition, disease, competition for resources, warfare, and even sociopolitical upheaval. But Haug stresses that "We can handle climate change if we're prepared for it. The Maya were not prepared."

The beginnings of civilization can be traced back to the development of the land through the conversion of wild lands into agricultural lands. Agriculture originated around 10,000 B.C.E. in at least five different places. Turkey and the Middle East began cultivating wheat, barley, peas, and lentils. People in these areas also raised sheep and goats. In Southeast Asia, people began to grow vegetables and raise pigs and chickens. In South America, agricultural development began independently in both the Andes and the Amazon regions. Northern China and West Africa also began their own development of agriculture.

Climate played a large part in making these farming areas productive. Temperature, amount of rainfall, and dependability of seasons all contributed to the success of supporting populations. Experts believe that the origin of agriculture and the spread of civilization are phenomena of a warm climate.

When historians look at past civilizations worldwide, they find a common thread linking them repeatedly. All of these great cultures assumed that good weather was guaranteed and would always continue—the climate would never change. Unfortunately, the recurrent pattern of history has been for civilizations to originate, grow and flourish while the climate is favorable; and then fall when the climate turns bad. This sense of false security has doomed civilizations time and again. And although other factors always play a part in determining the success or failure of a society, such as internal conflict, war, trade and economics, and other social factors, the droughts, famines, and pestilences caused by climate change take a predominant role in the survival of most societies, such as the Maya, Anasazi, Mesopotamia, Akkadian empire, Vikings, and others. According to Harvey Weiss, professor of archaeology at Yale University, "The historical lesson . . . is that those societies had no knowledge of what was happening to them and certainly no historic knowledge of what could happen to them, where we have both."

CENTRAL AND NORTH AMERICA

Some of the most notable past civilizations that have been directly influenced by climate are the Maya of Central America and the

Anasazi of the American Southwest. The Maya (also referred to as Mayans) were Central-American Indians. They were one of the greatest civilizations of the Western Hemisphere. Advanced for their time, they were well known for their extensive practice of agriculture, construction of enormous stone buildings and pyramid temples, artistic smithing of gold and copper, and development of a mathematics and hieroglyphic writing system. At its most advanced stage, Mayan civilization consisted of more than 40 cities, each populated with up to 50,000 people.

Today, all that is left of this impressive, sophisticated civilization are empty ruins deep within the wild jungles of southern Mexico, Guatemala, Belize, El Salvador, and northern Honduras. The civilization that thrived for 3,500 years (2000 B.C.E. to 1500 C.E.) abruptly disappeared. Just as Mayan society was reaching its height in 250 to 830 C.E., huge cities were suddenly abandoned, and the inhabitants seemed to vanish overnight. For years, historians pondered the reason for the sudden collapse of one of the greatest civilizations of all time. Then climatologists began to study sediment cores extracted from area lakes, and their findings offered clues to the mysterious disappearance of the Mayan civilization. Gill Richardson, an archaeologist who has spent considerable time studying the Maya, says that lack of water was a major factor in the collapse of the society. His most convincing evidence centers on sediment cores from several of Yucatan lakes acquired by David A. Hodell, Jason H. Curtis, and Mark Brenner from the University of Florida. Their measurements of the ancient deposits indicate that the driest interval of the last 7,000 years was between 800 and 1000 C.E. This coincides with the collapse of classic Maya civilization.

The Maya occupied the area known today as the Yucatán Peninsula, which has a seasonal desert climate and relies on annual rainfall for freshwater. Availability of abundant precipitation was important to the Maya because much of the area is underlain by limestone, a structure called karst topography. Karst topography erodes easily, so as rainwater collects in streams, it wears away the limestone beneath the ground and carves out caves, channels, and sinkholes. Although there were areas of abundant groundwater as a result, surface streams were not plentiful, which meant that rainwater for these people was very important.

Climate change played a role in the destruction of once-thriving Mayan civilizations. *(Nature's Images)*

Archaeologists have determined that because of this, the Maya developed extensive reservoir systems to capture, store, and distribute water. They were also very adept and efficient at agricultural practices and were able to make the desert productive. The more successful they were, the faster their population grew, until it reached a maximum of 200 people per 0.6 square mile (1 km²), one of the highest population densities of any ancient civilization.

Climatologists have retrieved sediment cores from both lakes and the ocean. What they found and were able to cross-verify from other sources put the Mayans' mysterious disappearance into perspective. According to experts at NOAA, the ratios of the oxygen isotopes ^{18}O and ^{16}O were analyzed in sediments taken from lakes in the Yucatán. These ratios were used to identify wet and dry periods in the climate. The oxygen isotopes were measured by the calcium carbonate ($CaCO_3$) found in the shells of the aquatic organisms deposited within the sediment cores. Drier climates are signified by higher amounts of the oxygen-18 isotope. Therefore, when the ^{18}O to ^{16}O ratio is higher, it tells climatologists the climate at the time was much drier and the area was probably experiencing a drought. Conversely, when the ^{18}O to ^{16}O ratio is lower, it signifies wetter conditions.

Climatologists discovered that shifts in atmospheric patterns caused a severe drought. Over an extended period, the Maya suffered several drought cycles, but from 800 to 1000, an extremely severe drought occurred in the region that devastated their ability to cultivate agriculture and maintain an adequate supply of water. During this time, the Mayan civilization collapsed. According to a 2003 article in National Geographic News, some archaeologists have estimated that the Maya were especially vulnerable to the climate because up to 95 percent of the population depended directly on lakes, ponds, and rivers for their supply of water, which contained only a year-and-a-half supply.

This conclusion was supported with evidence collected from ocean core sediments retrieved off the coast of Venezuela. Trace deposits of titanium, which are an indicator of weathering and erosion, were analyzed. High amounts indicate heavy weathering and erosion, which would indicate a wetter climate. What scientists found, however, were low amounts of titanium, indicating a drier climate. By dating the sediments, they were able to determine that not only was the area extremely dry, but there were several even more severe minidrought cycles within the longer period of drought. Dating techniques placed these minidrought cycles at 810, 860, and 910, which led scientists to believe that rapid climate change caused the rapid decline of Mayan civilization.

Also supporting this conclusion are data received from ice cores obtained in Greenland. During the same period, the ice cores had higher levels of ammonium, which indicates that the entire Northern Hemisphere experienced markedly drier climatic conditions during this period. According to NOAA scientists, the Mayan civilization faced a drought that lasted more than a century, and when their groundwater sources were depleted, their society collapsed.

Other cultures of Central America also met similar fates. The Teotihuacán, who lived north of present-day Mexico City, had an advanced civilization that fell around 800 due to deforestation of the surrounding

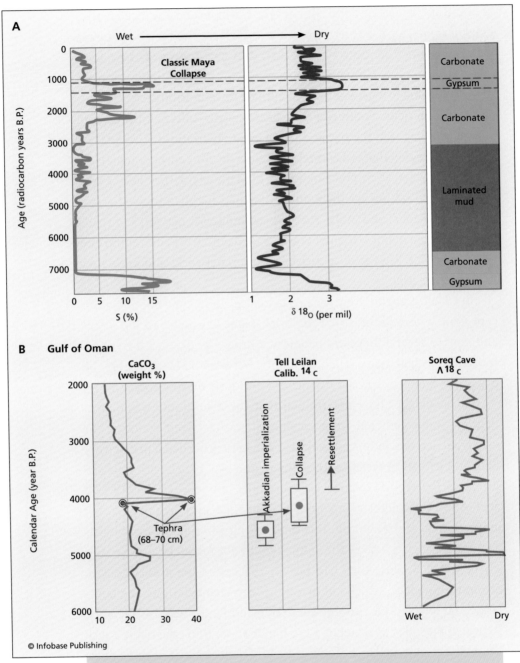

These two time lines illustrate that the collapse of two prominent civilizations—the Mayan and Akkadian—were related to severe climatic changes, including the abrupt onset of drought. *(Source: Hodell, et al., 1995, Nature)*

area and an increasingly arid climate. The Cacaxtla and Xochicalco, who lived near present-day Cuernavaca, also collapsed, and their settlements were abandoned around 800 because drought destroyed their food and water supplies. Climatologists believe this general trend occurred regionwide due to a general shift in climate toward aridity.

The American Southwest has been greatly affected by climate change and has seen the rise and fall of great civilizations. The Anasazi lived in the region that today is referred to as the four corners area (New Mexico, Arizona, Utah, and Colorado). Like the Maya, they were a very advanced civilization for their time. They had highly advanced astronomical calendars and tools. The Anasazi built vast networks of roads that connected major settlements and traded with tribes in Central and South America. They had extensive irrigation networks and knew how to farm efficiently.

In Chaco Canyon, New Mexico, and elsewhere, they built magnificent stone houses of sun-dried clay brick, some of which had 500 rooms and housed several thousand people at a time. The Anasazi thrived for more than 1,000 years. Then suddenly, between 1275 and 1300, all expansion stopped, and the Anasazi mysteriously disappeared.

Archaeologists have determined that the community was dependent on water gathered from ephemeral washes. They collected the water and diverted it through canals to where they needed it in extensive dam and irrigation systems. Then the region was struck with repeated droughts. Between 1125 and 1180, very little rain reached the dry, arid area. Another drought struck from 1270 to 1274, and then in 1275, yet another cycle of drought hit, this time lasting 14 years. Climatologists and archaeologists believe the culture could not maintain a stable supply of water. Because they had heavily deforested the area, extensive erosion took place, which cut arroyos (deep channels) into the landscape and lowered the groundwater table and hence water availability. Repeated episodes of drought were harsh enough to eventually cause the civilization to collapse. Today, the only remnants left of these magnificent settlements are a maze of stone ruins, a reminder of what climate change can do.

The Hohokam (their tribal name means "the people who vanished") were the ancestors of the present-day Pima Indians. Also living in the

Ruins found in Chaco Canyon left by the Anasazi are all that remain today of their great civilization that housed thousands of people. Long droughts contributed to the failure of their society 1,000 years ago. *(National Park Service)*

American Southwest, they were dwellers of the Salt-Gila River Basin in Arizona. Their civilization began around 300 B.C.E. and lasted until around 1450 C.E. To deal with the arid climate in which they lived, they built extensive networks of canals and irrigation systems. Then, beginning around 1080 and lasting for the next few hundred years, the area was repeatedly flooded, interspersed with long periods of drought. Their water supplies became contaminated with salt and dried out, spelling their eventual destruction.

There are also other signs of megadroughts throughout North America (900–1400 C.E.), such as massive sand dunes that are currently grassed over and stable. Much of the evidence found by climatologists has been derived from tree ring proxy records.

There have also been more recent examples of the repercussions of climate change. In the 1930s, the United States endured an extremely destructive environmental event during what is referred to as the dust bowl era. Because of improper farming practices, the topsoil was left damaged, dried out, and vulnerable to erosion. Unfortunately, the climate was in a drought phase at the time, and great amounts of silt, soil, and sand were blown hundreds and thousands of miles away, leaving the area infertile and uninhabitable. Farmers chose to abandon their farms and seek refuge elsewhere, such as California. It represented the largest migration in U.S. history.

Today, with ever-increasing population and settlement of communities in the Southwest, experts caution against rapid growth because of increased aridity predicted with an increase in global warming. Paleoclimatologists also warn that global climate is moving toward conditions that have not been seen for at least 10,000 years. To support this opinion, they note that the upward trend in CO_2 and temperature in the last half century points directly to a strong human contribution because of heavy fossil fuel use, deforestation, and poor farming practices. Some scientists believe that climate itself may not be the sole reason for a society to collapse but that it could act as one of a multitude of reasons. For instance, climate could cause a shortage of food and water, which could then destroy a country's security, generate social unrest, lead to war, and cause a civilization to fail. There are many combinations of possibilities, but scientists agree that climate is a principal factor in determining a society's success. They stress that climate must be looked at as multidimensional, influenced by both geophysical and social components.

MIDDLE EAST, FAR EAST, AND EUROPE

Abrupt climate change, droughts, cold periods, and flooding have also played a major role in other areas of the world. In the Middle East, the harsh arid climate has often been an issue, and some believe it may partially explain some of the long-standing animosities among peoples there.

People lived in northern Mesopotamia relying on annual rainfall in order to farm and grow their food for more than 1,000 years, and their society flourished. Archaeologists have found evidence, however, that roughly 8,000 years ago the inhabitants of northern Mesopotamia suddenly migrated to southern Mesopotamia. What struck archaeologists as odd initially was the fact that the southern area is much less hospitable for growing crops and survival. But as scientists began to dig for clues to these answers, they discovered, through the use of proxy data such as sediments and ice cores, that a major shift in climate occurred around 6300 B.C.E., which matches the period that the occupants of northern Mesopotamia migrated 8,000 years ago. Evidence has led scientists to believe that with the abrupt shift in climate in the northern region, annual precipitation necessary for farming may have decreased by nearly half, making the area uninhabitable.

The Akkadian empire ruled in the same area from the Tigris and Euphrates Rivers to the Persian Gulf about 3000 B.C.E., then collapsed about 2200 B.C.E. because the area became more arid. Climatologists have uncovered proxy data that support the climate change in sediment cores from the Gulf of Oman. Because the gulf is directly downwind of where the Akkadian empire was situated, as the area became drier, more dust was eroded and carried away by the wind and deposited in the water, where it collected in the sediments at the bottom. Through the analysis of sediment cores, scientists have determined that sudden climate change was a large factor in this society's collapse. The same trends occurred with the Nile civilization from 2181 to 2040 B.C.E., the Babylonian kingdom from 2200 to 1900 B.C.E., and the Indus civilization around 1800 B.C.E.

Likewise, when favorable climate conditions existed in Egypt and west Asia, the civilizations there thrived. Then, just after their peak in 2300 B.C.E., long-term severe drought struck, and cooling disrupted agricultural production in the area, which caused a famine that lasted for 200 years and brought about the collapse of these civilizations as well. The fall of early Bronze Age civilizations in modern-day Greece and India have also been linked to abrupt climate changes about 2200 B.C.E.

Archaeologists have determined that the ancient civilization in western India was also greatly influenced by climate. The paleoclimatic record has been determined through the use of microfossils as proxies to show that there was a minor drought that occurred around 2000

B.C.E., followed by a major drought around 1500 B.C.E., both of which seriously impacted the civilization in that portion of the world.

A similar sequence of events occurred in China. Between 700 and 900 C.E., the climate changed. Winter monsoons strengthened, but summer monsoons weakened, which contributed to cooler temperatures and shorter growing seasons. Many climatologists have concluded that the prolonged droughts that occurred caused massive crop failures and civil unrest, which led to the eventual collapse of the Tang dynasty. This occurred at roughly the same time as the demise of the great Mayan civilization.

The Vikings began to occupy Iceland around 870 C.E. At that time, Iceland's climate was much warmer and milder than it is today, making it possible to farm the land in a previously unsettled area. Their settlements flourished until approximately 1300 to 1400 C.E., when climatologists believe the climate abruptly changed. Climatologists have estimated that for each degree drop in average summer temperature, there was a 15 percent drop in crop yield. At that point, it turned cold, and the farms failed, animals died, and the settlers were left without food or supplies. It is believed that those who lived there at the time starved to death.

Archaeologists have been excavating sites in Iceland for the past century trying to gather clues about the effect abrupt climate change had on these early settlers. Adolf Fridriksson of the Icelandic Institute of Archaeology believes the land was covered with birch forests and that temperatures were very favorable when the Vikings first landed. He also points out that the area harbored fertile fishing grounds around the island and hosted rivers with plentiful trout and salmon, conditions ideal for settlement.

Fridriksson also points out that in addition to the change in climate, the Vikings also heavily deforested the island, another factor that led to the fall of their civilization. Like global warming, deforestation is a serious problem that compounds the problems of erosion and the natural balance of CO_2, which influences the temperature of the atmosphere and the overall cycle of the hydrosphere-biosphere-atmosphere system. Today, the inhabitants of Iceland say the climate is noticeably warming once again and that birch trees are returning. Climatologists believe this may be another testament to global warming.

The Bronze Age Argaric people of southeast Spain were another culture affected by climate change. Their civilization disappeared about 1600 B.C.E. In this case, archaeologists believe that climate played a partial role in the demise of the culture. Through analysis of pollen in sediments, they have been able to determine that the climate shifted to much drier conditions around 3500 B.C.E., indicated by evidence such as a reduction in forest cover, the appearance of plants adapted to drier climates, and a drop in regional lake levels. Added to this, however, was a human component. Settlers of the area also deforested much of the land, degrading the soil and vegetation, which ultimately led to their demise. This is an example of how the effect of poor land management practices can be magnified and harm inhabitants of a region.

ADAPTATIONS TODAY

While climate may not be the sole cause of failures of past civilizations (issues such as political unrest and warfare also cause collapses), it has repeatedly been, if not the main contributor, at least a key contributor to the stability of a society. Because of this, many researchers believe that humans today can learn much from the experiences of past civilizations. History is also helping scientists improve predictions through modeling and other developing technologies. But even with all of today's new innovations, there is still uncertainty about what the future will bring. The climate system is so complex there is still much research that needs to be completed.

One component humans must face today and that has not changed is the ability to adapt under changing conditions. If humans cannot adapt, they will perish. An example that illustrates this is when the Vikings inhabited Greenland during the Little Ice Age, which lasted from 1400 to 1800 C.E., and which put too much stress on the agricultural systems, forcing the Vikings to abandon their farms. At the same time, however, the Inuit, who lived nearby, continued on as they always had because they had already adapted to a colder climate. This focus on societies' responses to changing conditions is important when it comes to ultimate survival.

In light of global warming today, adaptation and the issue of lifestyle changes are also important. As reported in a National Geographic

News article in 2001, according to Wil Burns, a senior fellow at the Center for Global Law & Policy at the Santa Clara University School of Law and an affiliate of the Pacific Institute for Studies in Development, Environment, and Security (both in California), if the world does not change its current activities, an 11-degree rise in temperature could ensure extremely detrimental effects on civilization. Because most of the world's people live as subsistence (growing only what they need to survive) or small-scale farmers, fluctuations in climate could be devastating to these agricultural societies. Even though technology has advanced, most civilizations today would be negatively affected by droughts or monsoons; enough so that food production could be halted. Scientists also point out that as the Earth becomes more populated, it makes it more difficult for people to relocate to other areas that have more favorable climates.

Because we have a better scientific understanding of climate today, many experts believe that knowledge needs to be used to minimize the negative effects of climate change on societies that currently face the greatest risk. Several areas around the world are already putting adaptation to the test. For example, in Niger, inhabitants are currently planting trees to lessen the effects of flooding and aridity in the future. A group of farmers in England are currently taking advantage of the continuing rise in local temperatures by planting olive groves to produce olive oil, grape vineyards to produce wine, and tea plants, something unheard of in England's climate before now.

In order to preserve botanic diversity in view of changing climate, some areas are establishing seed banks. These are storage facilities that preserve the seeds of thousands of threatened and endangered species. Seed banks exist in the United States, Brazil, Mexico, England, South Korea, Norway, China, Russia, and Iraq.

By collecting and freezing seeds, scientists are trying to preserve the role that a particular plant plays in an ecosystem as well as its genetic diversity. They are also interested in preserving the seeds for their DNA so that future cross-breeding may be possible in order to produce plants that will be better able to grow in a changed climate. Seed banks are often looked at as a defense against climate change and environmental catastrophe.

Some critics of the theory of global warming say that some areas of the world may actually benefit from a warmer environment. This can be seen in a few places today. For example, in some parts of Australia, climate change has brought more rain, enabling some ranchers to support larger herds of cattle because their ranges have been more productive growing grass, and the farmers are benefitting from the extra productivity. Other northern locations in Canada and Russia will be better able to produce agricultural goods, but climatologists warn that areas of benefit will be few, far between, and sporadic and will be insignificant in comparison to the overall negative effects of global warming.

Like civilizations of the past, civilizations today are also living through the effects of changing climates. In 1995, for instance, an intense heat wave in Chicago, Illinois, killed more than 700 people and focused U.S. attention on climate change. In 2003, a record heat wave in Europe killed 49,000 people. Hurricanes Andrew in 1992 and Katrina in 2005 caused horrendous damage, as did the flooding of China's Yangtze River basin in 1998. In Zimbabwe in an El Niño year, corn yields experienced about a 10 percent loss. In Rwanda, temperature and rainfall extremes during El Niño caused increases in malaria outbreaks.

In the United States, State Farm Insurance has raised rates for inhabitants of Florida because of unpredictable and dangerous weather. Rates for homes have increased 70 percent and 95 percent for mobile homes. Southern California is facing a serious shortage of water, which will ultimately hurt the housing industry, the job market, and the local economy. Drought in the West is already overloading hydroelectric power production. Power shortages could reach the Pacific Northwest if the region's river flows drop below levels necessary to cool coal- and gas-fired power plants. Across the United States, climate change will disrupt jobs and the economy and will lead to health issues.

Global warming is affecting civilization today, and some areas are responding to these wake-up calls. According to the Earth Policy Institute, political leaders in sub-Saharan Africa are considering planting a 3-mile (5-km)-wide and 4,350-mile (7,000-km)-long belt of trees across the continent on the edge of the desert to stop its continual advance. They know that if they cannot build this Great Green Wall there will be millions of refugees left homeless as their lands steadily turn into desert.

In the Sahara in Mali near Timbuktu, the Tuareg people have traditionally lived as nomads, herding their animals from field to field. When they need to buy provisions at a market, they trade animals for other staples. Over the past 40 years, however, their traditional lifestyle has been disrupted because of persistent drought. In order to survive today, they have had to change their culture, settle in villages, and begin farming crops that are as resistant as possible to drought.

According to CBS News London in June 2007, another issue is the destruction of modern-day cultures that is occurring because of global warming and climate change. For instance, rising sea levels are currently encroaching on and destroying the ancient buildings of Venice, Italy, and the ancient temples of Greece. Higher temperatures and humidity are speeding up corrosion rates on the Eiffel Tower's steel girders in Paris. Humid, salty air is beginning to ruin the stonework of historical structures such as the Parthenon, the Tower of London, and many cathedrals throughout Europe. Scientists warn that many works of art such as these may not be recognizable or even exist in 50 to 100 years.

Many feel these wake-up calls need to be addressed and that action needs to be taken immediately to stave off the negative effects of climate change. Experts warn that climate change along with population growth, poverty, increasing water shortages, rising oil prices, and a potential rise in food prices could lead to political instability, which could result in wars. Because of this, global warming is an issue that today's politicians as well as scientists urgently need to address.

Using New Technology to Discover the Past

Continual advances in technology have made it possible for scientists to discover new clues about what the Earth's climate was once like. Two methods, remote sensing and geographic information systems, have proved particularly helpful. Scientists can now locate areas where major rivers once existed that are now arid desert, areas that once supported large lakes, and areas where a different climate fostered now-extinct ecosystems. This chapter looks at the field of remote sensing and how scientists use it to understand more of the climate history as well as geographic information systems (GIS) to classify, store, handle, and analyze all the data being collected today.

THE SCIENCE OF REMOTE SENSING

Remote sensing is the collection and measurement of information by a device not in physical contact with what it is observing. Common remote sensing devices include eyes, cameras, binoculars, microscopes, telescopes, video cameras, and satellites. When a 35-mm camera takes a photograph, for instance, a hard-copy print of the object is the output. If the picture were of a house, the photo interpreter would gain useful information such as the shape of the house, number of floors, number of windows, color, and landscaping. Using the photo, an interpreter is able to gather all this information without ever physically touching the house.

Remote sensing devices allow an interpreter to see objects at a distance or to see small objects better. Earth scientists use remote sensing to gather information about the Earth. The photographs they use are images that can be obtained from many types of remote sensing devices and that offer unique views of the Earth's surface unattainable in any other way. The most common platforms today are airplanes and satellites.

Satellites allow the Earth's surface, atmosphere, and oceans to be observed from space. Similar to humans, satellites have sensors that serve as their eyes. Satellites can see better than humans, however, because their sensors can detect much more of the Sun's electromagnetic energy, that is, all the energy that comes from the Sun. This energy travels through space from the Sun to the Earth and is composed of several ranges of wavelengths.

The shortest wavelengths in the spectrum are gamma rays, X-rays, and ultraviolet rays. Humans cannot see this energy (X-rays are what doctors use to take pictures of bones, and ultraviolet rays are what cause sunburn). As the wavelengths lengthens, visible light appears. These are the wavelengths that humans can see, but it is a tiny portion of the entire spectrum. Visible light can be broken into blue, green, and red light. Wavelengths longer than those of the visible spectrum cannot be seen by humans. These include infrared radiation, microwave radiation, and radio wavelengths. Scientists refer to these groups as bands (blue band, green band, red band, infrared band, microwave band, etc.).

Even though humans cannot see this energy, remote sensing imaging systems can. The various bands of wavelengths can be used to see different things. Most satellites can see objects in several different bands, and each band gives an image interpreter different information about the landscape. Some remote sensors can detect and record more than

200 different spectral bands. These types of sensors create "hyperspectral" imagery and are very useful to geologists looking for specific mineral deposits. Such bands are very narrow and highly discriminating.

Equally important as the wavelengths in which the remote sensor can "see" is image resolution, remote sensing system's ability to record and display fine detail. The smaller the resolution, the more detailed the image. When a digital image is acquired, it consists of rows and columns of numeric data. Each space on the resultant grid is a cell called a pixel (short for picture element). There is one numeric value for each pixel in each band, the size of which determines its resolution. If a satellite image has a 98-foot (30-m) pixel resolution, the smallest object in the image directly observable is at least 98 feet (30 m) in size. Different sensors on satellites have different pixel resolutions. The type of resolution scientists use for a project depends on what they need to see. If they wish to study large areas in general detail, then images with larger pixels work well. If they need to see specific detailed imagery, an image with smaller, more detailed pixels works well.

Computers are able to interpret the data contained in the imagery using sophisticated image analysis and classification software that sees the images as different spectral bands and looks at the numerical values of each pixel in each band to classify the image according to those patterns of numerical values. The computer can discriminate much better than the human eye, which can differentiate only a limited number of colors or tones. A computer can differentiate hundreds of values.

Image processing software performs many diverse functions. It can correct geometric properties so that an image better represents the ground it depicts. It can enhance, evaluate, and identify features based on principles of contrast and texture. It can display images in multiple classes based on multispectral ranges. It can look at different combinations of spectral bands to show highly diverse information and create data to be used.

Two basic types of remote sensing systems are used, active and passive. Passive systems record the energy emitted from the Earth. LANDSAT, QuickBird, and weather satellites are examples of passive systems—they merely record what they see reflected from the Earth's surface. Active systems generate their own energy, send out signals, and then record how they interact with the surface of the Earth. Active systems can operate from both aircraft and satellites.

An example of an active system is radar, a distinct form of remote sensing imagery. It transmits a microwave signal and then receives its reflection as the basis for forming digital or pictorial images of the Earth's surface. The radar contains a transmitter that sends repetitive pulses of microwave energy at a given frequency. It also has a receiver that accepts the reflected signal received by the antenna, then filters and amplifies it. It is the antenna array that transmits the narrow beam of microwave energy. Finally, a recorder logs and displays the signal as an image. Radar is the same technology as the radar guns highway patrol officers use to check a driver's speed.

Earth scientists use remote sensing for a wide variety of applications. It is useful for geologic applications, such as identifying the physical and chemical properties of rocks, understanding the relationships between plant cover at the Earth's surface and the structure of underlying rock, and studying faults, drainage systems, coastlines, and mountain systems. It can help hydrologists study water bodies, rivers, and streams and understand environmental effects from things such as drought, pollution, and global warming.

Remote sensing can also help archaeologists locate ancient ruins, which give paleoclimatologists clues to past climatic conditions. Remote sensors can detect irrigation ditches filled with sediment because they hold more moisture and have a different temperature than the surrounding soil. The ground above a buried stone wall may be slightly hotter than the surrounding terrain because the stone absorbs more heat. Radar waves can actually penetrate the ground to see what is under the soil.

Remote sensing can also be used as a discovery technique, since a computer can be programmed to look for distinctive signatures of energy emitted by a known site or feature in areas where surveys have not been conducted. Such signatures serve as recognition features, like fingerprints. Characteristics such as elevation, distance from water, distance between sites or cities, corridors, and transportation routes can help predict the location of potential archaeological sites. There are several examples of how remote sensing has been used to make archaeological and paleoclimatological discoveries:

- The Maya causeway (old remnants of trade routes used by the ancient Mayan culture) was detected through analysis of wavelengths in the infrared portion of the electromagnetic spectrum. Even though old paths may not be visible to the human eye, they appear different from vegetation in the infrared portion of the electromagnetic spectrum. Because they are discernable with imaging equipment, these ancient routes can be discovered, mapped, analyzed, and explored.
- A laser device called LIDAR (light detection and ranging) is a remote sensing system primarily used to collect topographic data that has been used to detect eroded footpaths that still affect the topography in many parts of the world. It is used extensively by NOAA and NASA to document subtle topographic changes in landscapes. LIDAR senses what is on the ground through using laser pulses similar to radar technology (which uses radio waves instead of light). LIDAR has proven highly successful in identifying archaeological sites.
- In 1982, radar from the space shuttle penetrated the sand of the Sudanese desert and revealed ancient watercourses. Because radar has the ability to sense beneath the surface of sand dunes, P-band (the 430-MHz wavelength range) microwave sensors have also detected ancient river drainage patterns, enabling researchers to link ancient settlement patterns with existing natural resources.

Some locations in the world where a better understanding of ancient climate has been gained through the use of remote sensing technology include Darfur in the Middle East, the Saraswati River in India, primeval forests in the United States, the Safsaf Oasis in Egypt, and the central Sahara in Africa.

Ancient Lake in Darfur, Middle East

Eman Ghoneim of the Boston University Center for Remote Sensing in Massachusetts was using radar satellite imagery taken of the northwestern Sudan to map ancient hydrology when she discovered evidence that an enormous lake once existed in the region—a sharp contrast to the very arid, parched environment that exists there today. This discovery confirmed that the climate in this part of the Sahara was once humid enough to support a green region with ample water and vegetation. Because radar waves have the capability to sense beneath the ground's surface, they were able to differentiate the subsurface layers and image them. What the radar revealed was a dark area roughly 0.6 miles (1 km) wide, in direct contrast to the bright spectral signatures of the sand around it. The low (dark) signature was caused by a mixture of sand and gravel, indicative of a shoreline from a long-gone ancient lake. In addition, Ghoneim identified nine lines that radiated from the shoreline, showing where rivers once drained into the ancient lake.

Further research confirmed that there was not just one shoreline, but four. This was important information because it told researchers that at one time the area had ample moisture and that the lake was fed by tributaries. Then, the supply of water began to lessen, and the lake began to shrink. At each subsequent shoreline, the lake spent some time in equilibrium until climatic conditions turned drier and caused the lake to shrink further. By dating these different shorelines, the area's climate history could be recreated and studied.

Ghoneim was able to combine the radar images with data from the Shuttle Radar Topography Mission (which provided landform elevation data) and create a model of the size of the lake. She determined the paleolake was pre-Holocene in age (older than 11,000 years) and was more than five times larger than Lake Erie.

Through further investigation, scientists have discovered that the water probably seeped through the sandstone beneath it and is now stored in deep groundwater reservoirs. Further proof of this comes from the fact that in 1953, Libya discovered underground reservoirs of water. This precious natural resource represents an important find for an area that will be hit extremely hard by global warming.

The Mysterious Saraswati River, India

Some 10,000 years ago, there were believed to be many mighty rivers that flowed from the Himalayas and that allowed civilizations to prosper in the green, fertile, cool climate on the riverbanks in northwestern India, such as Rajasthan. Archaeologists have determined that ample precipitation and large flowing rivers enabled settlers to be prosperous farmers. Then, 6,000 years ago, one of the mightiest rivers, the Saraswati, dried up, forcing inhabitants in the area to relocate elsewhere.

Over time, the Saraswati River, spoken of in ancient holy writings, slowly became a folklore legend because no one could find any physical trace of it. Then, through the use of satellite imagery, scientists recently discovered evidence of a once-major river 5 miles (8 km) wide that flowed through northwestern India. They were also able to determine that it dried up 4,000 years ago—the same time the Saraswati disappeared. Both climate and geology are believed to have ultimately caused its disappearance.

Scientists using remote sensing are currently working with India's water experts to drill boreholes to seek water under the desert in the same area. Water they have retrieved so far from deep under the riverbed has been carbon dated at about 4,000 years old. More than 1,000 archaeological sites have also been discovered along the river course. Many believe this old riverbed may be the ancient Saraswati. As in Darfur, the location of groundwater from ancient water courses could help large populations in the face of drought conditions from global warming.

Primeval Forests in the United States

Scientists in the United States have used remote sensing images from around the country to identify and inventory stands of existing primeval forest. Most of these old trees are found in rugged, steep, out-of-the-way areas that are undesirable to build in, farm on, or harvest for lumber. Paleoclimatologists are extremely interested in these areas because some of the trees are thousands of years old and can provide a valuable record of droughts and floods that have occurred throughout time, enabling them to better understand environmental issues such as global warming. To date, old-growth stands have been found in the New England states, the Carolinas, Arizona, California, Texas, Nevada, Oklahoma, and Virginia. Scientists are hopeful that a better understanding of the wet and dry cycles of climate will increase their knowledge of global warming, climate change, and long-term predictions.

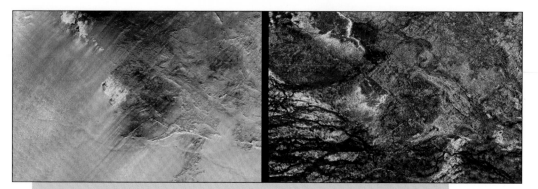

This image pair depicts the Safsaf Oasis in Egypt. The image on the left is produced from the LANDSAT satellite of the Earth's surface, showing very few drainagelike features. The image on the right is a radar image, depicting the subsurface rock features. The dark veinlike patterns depict ancient watercourses. *(NASA/JPL)*

Safsaf Oasis, Egypt

Visions of the Sahara today do not invoke images of major rivers and connecting multiple tributaries. Today the landscape looks like vast, open areas of nothing but sand for miles in all directions. As in the previous examples, even though the surface may not show any signs of water, this does not mean that water did not flow and cut tributaries at one time, only to later dry up and become covered with sand. At one time, abundant rivers in the Safsaf, Egypt, carved channels through canyons and formed lakes, and today these "fossil rivers" are buried under the sand.

In the photograph, the image on the left was obtained from a LANDSAT satellite. The surface looks hard and smooth, but upon close inspection one can see a very faint river channel that runs across the image. The image on the right of the same area, however, is what the subsurface of the ground looks like. This image was obtained from the Spaceborne Imaging Radar-C/X-band Synthetic Aperture Radar (SIR-C/X-SAR), which uses radar to penetrate the thin sand cover on the Earth's surface, and was taken aboard the NASA space shuttle *Endeavour*. This image shows that the area was once very different—the oasis was once a very productive, lush river valley. The sinuous dark channels (especially on the lower left) were cut by a meandering river and its tributaries. Such evidence helps climatologists reconstruct a region's climate in order to better understand the past, present, and future.

Central Sahara, Africa

NASA acquired a moderate-resolution imaging spectroradiometer (MODIS) image of the Sahara in Africa just north of Algeria and Libya. They studied three major rock formations among the reddish sand dunes: the Tassili, Tadrart-Acacus, and Amsak. Remote sensing specialists were able to identify several ancient riverbed structures in the Acacus and Amsak regions that followed a dendritic (treelike) pattern. Paleoclimatologists have interpreted this image and determined that the area was wet during the last glacial era, covered with forests, and probably inhabited by several species of animals. In addition, several renditions of ancient rock art have been found in the area, which indicates that the area provided a home for an ancient civilization. Experts have determined that the area became extremely arid about 3000 B.C.E.

THE SCIENCE OF GEOGRAPHIC INFORMATION SYSTEMS

A GIS is a computer system capable of capturing, storing, analyzing, and displaying geographically referenced information. A GIS operator can look at various types of data for a specific area and analyze them in order to make intelligent decisions.

A GIS can be used for scientific studies, resource management, and development planning. For example, a GIS might assist climatologists in determining areas where sea level rise will be the most disruptive, where desertification is spreading, how much impact El Niño is having, and the spatial relationships between drinking water, heat waves, drought, and disease.

The power of a GIS comes from its ability to relate different information in a spatial context and to reach a conclusion about that relationship. Most geographic information has a locational reference, which makes it spatial; that is, it fits somewhere on the Earth.

When data are collected by field scientists, they can be mapped or put in some kind of spatial reference system, such as a latitude and longitude coordinate system or any other map projection. When diverse data are all projected into the same system, that data can be overlaid and the layers analyzed in relation to other data concerning the same parcel of land in order to identify conflicts, problems, patterns, and relationships. For example, if a climatologist were interested in tracking the effects of global warming, all the pertinent spatial data involved could be reviewed, such as availability of water, shortage of food, vegetation cover, rainfall amounts, temperature patterns, disease outbreaks, ecosystem change, population densities, and natural resource inventories.

A GIS can convert existing digital information, which may not yet be in map form, into forms it can recognize and use. Digital satellite images can be analyzed to produce a map of digital information about land use and land cover. A GIS can also convert tabular data into a map-like form that a GIS can recognize.

There are several ways to enter data into an operable GIS system. Features on existing paper maps can be digitized by hand-tracing with a computer mouse on a computer screen or on a digitizing tablet to collect the coordinates of features. Scanners can convert maps to a digital product. Data collected from a global positioning system (GPS) can also be incorporated into a GIS. Data capture involves identifying objects on a map, their coordinate locations on the Earth's surface, and their spatial relationships. Many software programs have been developed to automatically extract features from remote sensing imagery.

Objects are identified in a series of attribute tables (the information part of a GIS). Spatial relationships, such as whether features intersect, are adjacent, or are separate from each other, are the key to all GIS-based analysis. This enables Earth scientists to look at many variables related to an area and study cause, effect, and interactions and allows scientists to better study variables that may not have obvious correlations. Looking at data on wildlife habitat and comparing it to the availability of food, existing habitat, and migration opportunities, for example, gives scientists insights about the way global warming may affect wildlife.

One of the key requirements of a GIS is the maintenance of data consistency and integrity. Data may originally be collected at different scales or in varied formats. The GIS must convert these data sets into a compatible format, such as a common map projection. Data must also be accurate. If data entered into a GIS is of high quality, then the output

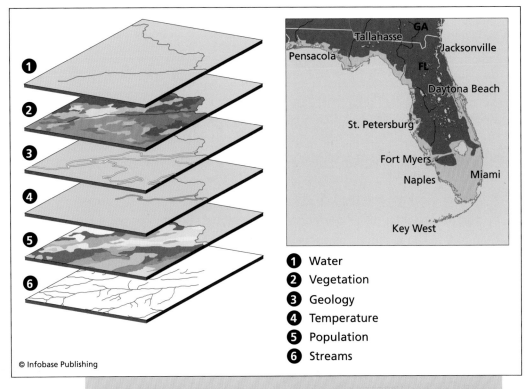

1 Water
2 Vegetation
3 Geology
4 Temperature
5 Population
6 Streams

© Infobase Publishing

Multiple layers of information can be stacked in a GIS to analyze how different components of the landscape affect each other. One GIS application is to model sea-level rise and its effects on major coastal areas worldwide.

from GIS analysis will be high quality; the final product is only as good as its least accurate input.

One tremendous achievement of GIS systems in recent years is their ability to model data to predict two- and three-dimensional characteristics of the Earth's surface, subsurface, and atmosphere, from which working models of cause and effect can be created. This real-life advantage allows scientists to check possible interactions and relationships of diverse phenomena and make intelligent decisions concerning them.

A GIS can recognize and analyze spatial relationships among mapped phenomena. Conditions of adjacency (what is next to what), containment (what is enclosed by what), and proximity (how close something is to something else) can be determined with a GIS.

A critical component of a GIS is its ability to produce graphics on a screen or on paper to convey the results of analysis. Wall maps, Internet maps, interactive maps, and other graphics can be generated to enable decision makers to visualize and understand the results of the analysis and simulations of potential events. The illustration on page 141 depicts the basic architecture of a GIS and how multiple layers can be stacked to make detailed analyses of the land and Earth systems as well as how an application to model sea-level rise can be used to determine how areas will be affected if ice caps and glaciers melt. In this example, the Florida Bureau of Geology and the USGS Geologic Division mapped the areas during the last interglacial period, when the global climate was warmer than today. The sea level then was 23 to 26 feet (7–8 m) higher than it is today. Climatologists believe these earlier sea levels may be reached again if global warming continues. Wildlife habitats such as the Everglades and other key features that affect the lives of more than 7 million residents of Florida would be negatively affected if this occurred. Using GIS, these impacts can be modeled and give not only scientists a clear view of the consequences of global warming, but also politicians, planners, environmentalists, and the population in general, showing why it is so important for everyone to become familiar with the realities of global warming and climate change.

What the Experts Say

Today scientists are researching different aspects of global warming: useful proxies, triggers, mechanisms, and other contributing and key factors. One of the leaders in research today is the U.S. National Aeronautics and Space Administration (NASA). This chapter presents some of its research concerning global warming and climate change.

MODELING ABRUPT CLIMATE CHANGE

In their efforts to predict future climate change and possible effects of global warming, scientists at NASA are looking at well-documented events from the past as "blueprints" for the future. They believe that by focusing on a past event and entering relevant data and associated proxy data into a computer model, if the model is developed correctly and its output matches the actual conditions that resulted in the past, then that working model can be used with today's observations to predict future scenarios. According to Gavin A. Schmidt, a researcher at

NASA's Goddard Institute for Space Studies (GISS), "We only have one example of how the climate reacts to changes: the past. If we're going to accurately simulate the Earth's future, we need to be able to replicate past events."

According to Allegra LeGrande and Gavin Schmidt, both scientists at NASA, varying specific components of a model can generate future scenarios. For instance, by varying the level of CO_2 in various iterations of a model, scientists can gain a clearer picture of what the Earth's environment would be like under those conditions. Likewise, a model can be run repeatedly and changes can be observed by varying the atmospheric temperature, the amount of precipitation, the flow of ocean currents, the percentage of ice cover, the percentage of freshwater, and any other variable related to global warming and climate change.

When designing such a model, scientists try to find a past event that lends itself to the creation of a solid, reliable model. To accomplish this, a model must have widespread and clear data, it must be an event that actually caused a change in the climate in the past, and it must be of a limited, defined duration. In line with these requirements, scientists at NASA decided to model abrupt climate change. The past event they chose on which to base the model was an abrupt cooling event that occurred across the Northern Hemisphere roughly 8,200 years ago. Fortunately, this event has been well documented with many different types of paleoclimate records as well as a significant geologic event—the catastrophic draining of the prehistoric glacial Lakes Agassiz and Ojibway in Canada. They both drained into the Hudson Bay at about the same time, flushing the North Atlantic with an overload of freshwater.

Climatologists believe this, in turn, disrupted the flow of the meridional overturning circulation (commonly referred to as the MOC) in the North Atlantic. Also referred to as the global conveyor belt, the MOC is critical for the distribution of warmth in the Atlantic. The current is an enormous continuous loop that brings warm water up from the equator to the polar regions. This warm water is what keeps western Europe's climate mild, considering how far north in latitude it is. When the current reaches its northernmost point and cools, it becomes dense, sinks, and begins its journey back toward the equator, where it is warmed again. The salinity level also plays an important role in the overturning of the current.

The problem with adding large amounts of freshwater to the ocean—by emptying a lake or through the melting of ice caps or glaciers—is that it decreases the salinity of the ocean water and slows the overturning process at high latitudes. Slowing the process slows the entire conveyor belt, which means that warmth from the equator is not brought into the Northern Hemisphere.

Using this information, NASA developed an ocean-atmosphere model that was analyzed by the Intergovernmental Panel on Climate Change (IPCC) in 2007 for their fourth assessment report. The model they developed, called a GISS ModelE, took the hypothesized cause (the draining of the lakes) and tried to reproduce the response. They accomplished this by adding calculated volumes of freshwater over predetermined intervals. To make the model as consistent as possible, they added tracers into the model, such as water isotopes, methane, dust, and other aerosols to account for climate proxies of the past, such as ice cores, cave records, and ocean and lake sediments.

Once the model was run, the scientists determined that the addition of the freshwater from the two ancient lakes to Hudson Bay was indeed enough input to slow down the conveyor belt current and cause a definite cooling in the Northern Hemisphere. In fact, it could slow the rate of flow of the MOC from 30 to 60 percent and cause a cooling of 3.5 to 5.5°F (2–3°C) in the North Atlantic as well as shifts in rainfall bands to the south in both the North Atlantic and the Tropics.

The results of this test are significant to the issue of global warming today and the idea of freshwater forcing. According to Allegra LeGrande and Gavin A. Schmidt of NASA, freshwater forcing could happen from the atmosphere heating and the ice caps and glaciers melting and draining into the ocean, ocean temperature warming, and increased rainfall.

"The flood we looked at was even larger than anything that could happen today," LeGrande reported. "Still, it's important for us to study because the real thing occurred during a period when conditions were not that much different from the present day." "Hopefully, successful simulations of the past such as this will increase confidence in the validity of model projections," added Gavin Schmidt.

MINERAL CLUES TO PAST CLIMATE

Like tree rings, ice cores, oxygen isotopes, and coral, minerals are another proxy used to reconstruct past climates. They can leave specific clues about not only the environment they were formed in but the ancient climate as well. Minerals react with both water and the atmosphere by being shaped and physically and chemically weathered. The most useful are those that exist within specific environmental settings, such as arid areas, tropical areas, and polar areas.

According to Dr. Vivien Gornitz of NASA, "Minerals furnish important clues about ancient climates. The most useful are those deposited under relatively narrow climatic ranges or within specific environmental settings." When geologists identify specific mineral deposits, the required conditions for their formation can tell a lot about the climate at the time of the formation. In arid areas, for instance, evaporite minerals are common. They form by the evaporation of seawater; evaporation of lakes in narrow, closed basins; and under very hot and dry conditions. If the minerals are in an enclosed basin (with no rivers flowing out), then as the water slowly evaporates, salts will begin to precipitate out in a certain order.

Beginning with carbonates, they precipitate to sulfates and end with chloride salts. By identifying the stage to which they are precipitated, assumptions can be made about the climate of the area. Common evaporite minerals include halite (rock salt), gypsum, borax, anhydrite, and nitrates. An example of a major evaporite deposit is the Bonneville Salt Flats located in the Great Basin province in western Utah. Located in a closed basin, the world-famous salt flats are a remnant of an ancient lake, Lake Bonneville, that existed during the last ice age. Gypsum and anhydrite are common, usually found in desert playas and arid coastal tidal pools.

Ancient sand dunes are other evidence of past arid climates. Fossil desert sands can contain information about wind direction and severity because sometimes wind ripples and cross-bedding evidence is preserved in fossil sand dunes.

Scientists at NASA also look for clues to past climates in soils and sediments. For example, the specific type of clay and its relative abundance is related to climate. In warm, humid climates in which chemical weathering is significant, silica is leeched out of the soil profile, leaving the soils with high concentrations of alumina.

Chlorite and illite, two other types of clay, are found in soils subjected to mechanical weathering processes. They can occur in both hot and dry climates and colder climates. By analyzing oxygen isotopes in clays sampled from ocean sediment cores, glaciation episodes can be studied.

When evidence of chemical weathering is present in soils and bauxite is the predominant mineral, it is evident that the climate at the time of formation was humid, wet, and tropical. These areas also have concentrated amounts of aluminum minerals such as iron oxides, kaolinite, and gibbsite.

NASA has also uncovered information about the Earth's ancient atmosphere. The clues lie in the fact that iron can occur in several oxidation states. Under the conditions of the Earth's present atmosphere, pyrite (fool's gold) oxidizes rapidly because of the abundance of oxygen. In ancient times, however, the atmosphere was not so oxygen rich, which caused pyrite to form well-rounded grains. These deposits have been found in sediments dated at 3 billion years old in the Witwatersrand basin in South Africa. This tells scientists that the Earth's atmosphere during the Archean eon (more than 2.5 billion years ago) had much lower oxygen levels than it has today.

Mineral deposits also reveal past cold climates. The presence of ikaite, a mineral that occurs at temperatures up to 45°F (7°C), indicates an environment in which near-freezing water temperatures existed at one time.

THE STORIES MARSHES TELL

Scientists at NASA and Columbia University have unraveled important pieces of the climate puzzle in the marshes of the lower Hudson Valley near New York City. Through analysis of sediment layers from tidal marshes in an estuary, Dee Peterson and Dorothy M. Peteet have discovered valuable proxy data in preserved pollen from various plants and seeds. Working with a team of scientists from Columbia University, they have uncovered physical evidence permanently recorded in these sediments of two past significant climatic events:

the Medieval Warm Period, a 500-year-long drought that spanned from 800 to 1300, and the Little Ice Age, the cold period dating from 1400 to 1850.

Through analysis of the pollen proxy in the Piermont Marsh sediments, they confirmed that the dominant vegetation during the Medieval Warm Period consisted of pine and hickory trees. There was also plentiful evidence of charcoal. This is significant because it is indicative of wildfires, another sign of drought. During periods of prolonged drought, vegetation dries out, leaving it vulnerable to lightning strikes. As more dried vegetation accumulates, fire danger increases, so that when lightning strikes it can cause extensive wildfires, leaving deposits of charcoal and ash that become buried in the sediments as a permanent record.

Also present in the sediment record dating earlier than pollen from the pine and hickory was evidence of oak, an indicator of a wetter climate before the Medieval Warm Period. Scientists see this paleorecord as an example of a natural fluctuation in climate rather than a human-caused change, such as those occurring today with global warming. Other evidence to support the occurrence of the Medieval Warm Period includes signs of erosion and an increase in salty marsh plant species in the bay (the drier an area, the more likely salt-resistant species will exist). Scientists view this as important information in relation to global warming today. If continued warming were to trigger drought conditions again in the area, salinity could cause serious problems with water quality and threaten the well-being of nearby populations through contamination of fresh drinking water.

Scientists were also able to discern the Little Ice Age through a change in dominant vegetation species to those that favored cooler and wetter climates. In this case, they found an abundance of spruce and hemlock, forest species that thrive in cooler, wetter climates.

The results of this study further support the use of proxy data to reconstruct the Earth's past climate in order to learn how changes affect the environment and the life that depends on it. The goal is to learn the lessons the data teach, apply them to the conditions that exist on Earth today in light of global warming, and make educated, wise decisions about the future.

WORLD WARMTH EDGING ANCIENT LEVELS

GISS scientists have determined that the Earth's current temperatures are now reaching a level that has not been reached in thousands of years. According to NASA global warming expert, Dr. James E. Hansen, based on a joint study conducted by NASA GISS, Columbia University, Sigma Space Partners, Inc., and the University of California at Santa Barbara, the Earth is surpassing the warmest levels it has seen since the last ice age, which ended about 12,000 years ago. Careful calculations have shown that for the past three decades, the Earth has been warming about 0.36°F (0.2°C) per decade. Although this may not seem to be a lot, it is. It is already causing many species of plants and animals to move toward the polar regions. They must do this because they can tolerate only certain temperature ranges. If their habitat becomes too warm, they need to move to a cooler area in order to stay in their comfort range, which means they must move poleward.

According to Dr. Hansen, head of GISS, "This evidence implies that we are getting close to dangerous levels of human-made (anthropogenic) pollution. In recent decades, human-made greenhouses (GHGs) have become the dominant climate change factor." One major obstacle, however, is that it may not be possible to migrate. For instance, animals cannot migrate if a city is in the way. Plants may not be able to migrate if the right kind of soil or moisture is not available. When one part of an ecosystem is disrupted, it causes a ripple effect and can destroy entire food chains and habitats.

According to Dr. Hansen, a study conducted in 2003 that appeared in *Nature* determined that 1,700 plant, animal, and insect species had moved poleward at a rate of 4 miles (6 km) each decade from 1950 to the present. With current global warming trends, however, the temperature zones have migrated 25 miles (40 km) per decade from 1975 to the present, meaning that species cannot keep up with the rate of change under global warming. Wildlife species that cannot keep up will face extinction.

The same threat exists for species that live in mountain habitats. When temperatures become warmer, species that live in the lower elevations may try to migrate to higher elevations, where it is cooler.

This may or may not be possible, especially for the species that already inhabited the highest elevations. They may be displaced with nowhere else to go.

According to Dr. Hansen, over the past 30 years, the increase in temperature—current global warming—is due primarily to human-made greenhouse gases. The area where the largest increases in temperature are occurring are the high latitudes of the Northern Hemisphere (the polar regions), and the temperature increase is greater over land than over the oceans. The reason polar areas are affected the most is due to the behavior of polar ice and snow. Snow and ice have a high albedo—they reflect most of the light they receive. As the Earth warms, however, it causes the snow and ice to melt, which exposes the darker surface of the ground beneath the ice. Because dark surfaces absorb light and convert it to heat energy, this causes accelerated melting of snow and ice in a chain reaction called positive feedback. The ocean does not heat as rapidly because water has a higher heat capacity—it can hold huge amounts of heat at extreme depths before it starts to warm significantly.

Based on research Dr. Hansen conducted with David Lea and Martin Medina-Elizade, they determined that the western equatorial Pacific and Indian Oceans are now as warm or warmer than at any time during the past 12,000 years. This conclusion was based on proxy data from the magnesium content found in the shells of microscopic sea-surface animals contained in ocean sediments. This is especially important because temperature change in these two oceans indicates overall global temperature change. The western Pacific serves as a major source of heat for the world's oceans and for the global atmosphere. In addition, the Earth's recent rapid warming has brought global temperatures to within 1.8°F (1°C) of the maximum estimated temperature during the past million years. In contrast, the eastern Pacific has not shown the same amount of warming. Hansen believes this is due to the effects of El Niño.

Based on the results of this study, Hansen concluded that if warming increases another 1.8°F (1°C), the Earth's atmosphere will reach a critical level. If temperatures stay below that level, the situation may remain manageable. If temperatures increase 3.3 to 5°F (2–3°C), however, the Earth will become a very different environment than what humans are presently used to. The last time the atmosphere was that warm (approximately 3 million years ago, during the Pliocene), sea level was 80 feet (25 m) higher than it is today. If that situation were to occur again today, it would flood coastal areas worldwide, causing serious disruption, damage, destruction, and death.

Conclusions and a Glance into the Future

Other than purely for the sake of knowledge, one of the key reasons scientists have such an interest in understanding the climate of the Earth's ancient past is so they can understand how the atmosphere and climate respond under varying conditions. The Earth's climate is an extremely complex system, and a change in any single component can have far-reaching effects on both a short- and long-term basis. For this reason, the more scientists understand about the past, the better they can predict the future. This chapter focuses on that objective—the creation of mathematical models as a way to accurately and realistically represent the climate. It also explores possibilities of what the climate on Earth may be like far into the future.

MODELING THE EARTH'S CLIMATE

Today, several scientific endeavors are attempting to model the Earth's weather and climate for a variety of reasons, such as for farming, urbanization, and emergency preparedness and for economic, scientific, political, and humanitarian reasons. GISS has taken a lead and become one of the premier groups involved in modeling climate in order to better understand it.

One of the main goals of the researchers is to be able to anticipate the effect climate change will have on society and the environment. Although they are involved with several types of models, they are currently focusing most on global climate models (GCMs). These are large-scale models with the ability to simulate the entire Earth and all the forces that affect it, both human-induced and natural. For example, natural forces include volcanic eruptions, variations in insolation (incoming solar radiation), and changes in the Earth's orbital path. Human-induced forces include pollution (increasing greenhouse gases from burning fossil fuels), adding aerosols to the atmosphere, ozone depletion, some types of farming practices, and deforestation.

When scientists create climate models, they strive to ensure that physical phenomena are presented as accurately and consistently as possible and that all components of the cycle are realistically receptive to changes in the system. This is not an easy task. Because so many variables are dependent on other variables, if something does not work well in a model, it can skew the results or cause the model to fail. Complex GCMs are validated when they reproduce exactly the results of an actual past climate response whose initial conditions were fed into the model as a test. In other words, a model is a success when it can accurately simulate changes that have already occurred.

Scientists desire to build models that can look back millions of years, which is why proxy data are so important. By using proxy data that portray ancient climate accurately, a model can be written and tested until the correct outcome is achieved. Once this happens and the model is validated, current data can be input to make projections of future climate. GISS researchers have already simulated many of the Earth's past climate periods to validate their models and to better understand the Earth's climate history. When they model global climate events, such as ice ages, they gain a clearer insight into the worldwide effects of global warming today.

Climate scientists at GISS, such as Gavin Schmidt, James Hansen, Allegra LeGrande, Drew Shindell, Nadine Unger, Leonard Druyan, and Matthew Fulakeza, have been highly successful in developing mathematical models that illustrate how changes on the Earth's surface and in the atmosphere are affecting the climate today. Because the climate is such a complex system, the models are highly sophisticated and complex, with countless variables that have to be taken into account. When even one variable undergoes the slightest change, it can affect many other variables in a domino effect, all of which must be taken into account and provided for through representative equations and algorithms. As an illustration, consider these four simple changes.

1. an area experiences less cloud cover over a given period and receives more direct sunlight
2. if the area was normally covered with snow, the snow could melt
3. the albedo of the region could change from higher to lower values because of snowmelt
4. the temperature of the area would increase

Just these changes for one area would have to be accounted for in the model, and this does not even address the other issues that would apply at that same site, such as humidity, slope, aspect, soil type, elevation, latitude, continental or coastal location, and other important aspects of that particular area.

Fortunately, within the last decade, computing power has increased tremendously. More than three decades ago, when NASA successfully put astronauts on the Moon, the computer that was used filled an entire room. In comparison, some of today's desktop computers are more powerful. Even so, the mathematical models and computing power necessary to run a GCM are enormous, even by today's standards, with computing power that has increased by a factor of a million in the past 30 years. Modeling systems must be able to handle a multitude of simulations simultaneously on both global and regional scales, taking into account characteristics on land, in water, and in the atmosphere. They must also be able to handle various timescales such as years, decades, centuries, millennia, and so forth in order to generate reliable scenarios of climate change. Supercomputers must also have significant storage capacity.

According to scientists at NASA, the most sophisticated models developed represent the Earth as a three-dimensional grid, with the atmosphere split into 10 different grid layers. Each one of these grid layers in itself is an enormous dataset containing 65,000 reference points. When the model is run, each of the 65,000 points has a data value associated with it that is used in the model. To make it even more difficult, each point has more than one data variable assigned to it, such as a value for CO_2, temperature, aerosol, pollution, albedo, and so on.

The power of the model lies in its ability to predict how the entire system will respond as values vary. For instance, if scientists want to see what a doubling of CO_2 will do, the model doubles CO_2 at all 65,000 points and predicts an outcome in relation to all the other variables. According to scientists at NASA, climate models are so intense that they have to be run on supercomputers that can handle more then 80 million calculations per hour. Simply to run a single model, the supercomputers must solve billions of calculations.

Even though the scientists at NASA have made great strides in modeling the climate, they still have not answered all the questions they would like to. According to Gavin Schmidt, a climate modeler at GISS, one thing they have not been able to model is abrupt climate change. He hopes that by combining paleoclimate and satellite image data NASA will eventually be able to build a model that will reveal aspects of the Earth's climate that have not yet been discovered and that will shed light on current climate mysteries. While he is interested in modeling paleoclimate, he says that "a further way to combine the models and the data is to 'forward model' the signal that would be recorded in the sediments or corals given a modeled climatic event."

In order to validate models, NASA put one to the test using the 1991 eruption of Mount Pinatubo in the Philippines. An extremely explosive eruption—one of the most violent in the 20th century—Mount Pinatubo ejected ash 21 miles (34 km) into the atmosphere. Scientists calculated that it also sent 17 million tons (15 million tonnes) of sulfur dioxide into the stratosphere. Because the stratosphere lies above the layer of the atmosphere in which Earth's weather takes place (5–31 miles, or 8–50 km), global atmospheric circulation spread the resulting sulfate aerosols (small reflective particles) around the Earth, so that after three weeks, the entire planet was encircled. As a result (rain could not wash them away), the resulting aerosol cloud remained in the stratosphere for more than a year, effectively shielding the Earth from the Sun's energy. This caused a measurable decrease in global temperature of 0.8°F (0.5°C). In addition, atmospheric water vapor decreased during this time. The end result was that the Earth measurably cooled for several years.

Because this eruption occurred at a time when satellite technology was available to observe and record it, climate modelers collected extensive amounts of data, such as sulfate measurements. Scientists hoped that if the recorded amounts they added to the models they had developed yielded results that matched those that had been collected in the field, it would prove the models were successful.

They found that the models generally predicted the results very well. The only problem was that they were not able to predict that Eurasia would warm slightly in the winter, as it actually did after the eruption. One explanation was that most global climate models do not usually deal with the stratosphere because it does not directly affect the weather. When the data were run on models that specifically included the stratosphere, however, they did achieve reliable results.

Scientists determined that Eurasia experienced a winter warming due to the North Atlantic oscillation (NAO) and that the sulfate cloud from Pinatubo had affected it. The NAO is a pressure system that determines how severe the winters in Europe are each year. A positive NAO makes Eurasia warm, and a negative NAO causes a cold winter. They determined the eruption triggered a positive NAO, which in turn made Eurasia warmer. To validate this, they ran similar related data that reflected conditions during the Maunder Minimum, which occurred from 1650 to 1710. By controlling the variable that represented the effect of the Sun's ultraviolet rays in the stratosphere and ozone production levels to match the conditions during that time, they showed that the NAO shifted to a negative phase and made Eurasia colder, which is actually what did happen.

Therefore, discovering the connection between the stratosphere and the NAO answered several complicated questions about how various climate factors interact with one another. Hence, paleoclimatology can be very useful by providing information, validating and refining models, seeing into the past, understanding better the ways that complicated climate systems work, and predicting what the future may bring both in terms of natural phenomena and human-induced effects.

Over the past 15 years, government researchers, private organizations, and academic institutions, in attempting to predict future climate change, have developed several global climate models. Each of these highly sophisticated 3-D models must have grids of cells programmed to solve for mass, momentum, and energy through timed sequences so that the climate system can be observed as it is modeled. The model is validated against observations and can also be run backward to see how much current models of a particular location match the actual behavior of a known site. Because there are so many input and output variables involved, they have to be continuously checked against experimental data to verify the results. Because of the importance of global warming, a common application of climate models today involves the effects of changing amounts of CO_2 in the atmosphere. Models have been run, for instance, to illustrate the potential effects of a doubling of the amount of CO_2, as is expected sometime within the next decade. If this were to happen, the results predict that the Earth will become much hotter and more humid and that sea level will increase 20 feet (6 m) over the next 100 years. The same models predict an increase of 6.7°F (4°C) over the same 100-year interval.

WHAT SOME MODELS SAY ABOUT NORTH AMERICA

Considerable ongoing research is aimed at predicting what the future climate will be like for North America in light of natural changes as well as anthropogenic (human-caused) global warming. Extreme drought is one of the expected consequences of increased global warming, especially in the American Southwest, where it has already been projected to be severe by several models.

The Drought Research Lab at Lamont-Doherty Earth Observatory at Columbia University in New York has identified and studied several

Drifting, blowing soil is burying a farm and its abandoned farm equipment in Kansas during the famous dust bowl in 1935. *(NOAA's National Weather Service Collection)*

drought cycles that have affected North America in the past. One of the most well-known and publicized droughts was the dust bowl of the 1930s, but it was not the worst. Through analysis of proxy data such as tree rings, scientists at Columbia were able to determine that in the late 1200s North America suffered a severe drought. This coincides with the period in which the Anasazi civilization in the Southwest disappeared (see chapter 7).

They also determined that during the entire Holocene (the past 10,000 years), most of North America has experienced periods of prolonged drought interspersed with wetter periods, called pluvials, and that some of the droughts were extreme. There was also what has been referred to as a megadrought in the late 1500s.

North America has experienced drought in recent times, as well. The mid-1850s to 1860s brought a severe drought to the Plains worse than the dust bowl. During the 1950s, much of the Southwest was dry. Then Arizona, New Mexico, Utah, Nevada, and California suffered a long dry spell from 1998 through 2004, and some of these areas continue to suffer today. Many of the reservoirs in these states are still well below their holding capacities, some of them 50 to 100 feet (15–30 m) low. Lake Powell, on the border of Utah and Arizona, for instance, is still so low that the original boat docks have been unusable for several years because the water no longer reaches them.

Low reservoir levels not only put a huge strain on drinking water supplies for millions of people who live in the Southwest but also harm farming practices, the grazing industry, and manufacturing industries in the area, all of which rely on a plentiful water supply in order to function at full capacity. Another major problem is the continually growing population. Many people are moving to the Southwest because they enjoy the mild winter climate, but as the population increases, so does the demand for freshwater. Unfortunately, much of the population growth in Arizona occurred from 1977 to 1998, which was an uncharacteristically wet period for the Southwest, meaning that construction and population boomed under atypical natural conditions, giving residents a false sense of security about water. If global warming is not controlled and the drought situation continues to worsen because of human activity, this will make survival in the desert Southwest even more challenging. In fact, the models climate scientists have developed predict that droughts will continue and become even more pronounced because of global warming.

Researchers have come up with two theories about why this area is affected by persistent drought. They believe it relates to subtle changes in sea surface temperatures (SSTs) of the tropical oceans, especially the tropical Pacific, and happens because of the way it causes the atmosphere to move. As hot air is warmed in the Tropics and circulates, it descends in the subtropical higher latitude areas. As this air descends over the American Southwest, it becomes very dry, bringing very little precipitation, which in turn causes drought. They have also noted that when La Niña events occur, they are usually associated with dry winters in the Southwest, which then causes the Plains states to have dry springs. If these dry episodes persist, the entire summer may remain droughtlike.

Another way researchers think the cold equatorial Pacific waters cause drought over North America is by relocating heat sources in the Tropics. If the Intertropical Convergence Zone (ITCZ), a convergence of warm, moist air at the equator, moves north where waters may be warmer, it changes global circulation, which in turn changes wind circulation. The net effect is an increase in the descending motion over western North America, which increases the drying effect and makes drought more pronounced.

One important point researchers have discovered at Columbia is that these shifts in water, atmospheric circulation, or temperature variation do not have to be extreme. It does not take a huge catastrophic event to cause a major change in the climate, such as drought. Instead, it is more likely to happen subtly, in small increments at a time. During the dust bowl and the 1950s droughts, for example, the tropical Pacific Ocean was only a few tenths of a degree Fahrenheit cooler.

According to Richard Seager at the Lamont-Doherty Earth Observatory, the important result of these models is that "persistent droughts are the sum effect of persistent, but small, precipitation anomalies. During the dust bowl precipitation over the Plains was reduced by about 15 percent but, when this happens year after year, the ground moisture gets less and less as evaporation proceeds, resulting in severe drought."

Seager also points out that persistent droughts do not affect only North America. When droughts occur, they occur in a global pattern of "precipitation variability," which means that worldwide some regions will also experience drought while others may experience increased precipitation. When North America experiences significant droughts, it is common for regions such as South America, the Mediterranean-Europe region, North Africa, the Middle East, and Asia also to have droughts, because all these regions lie in the global mid-latitudes (the subtropical regions) and are affected by the descending dry air in the atmospheric circulation.

Climatologists very much want to be able to predict when droughts and pluvial periods will occur. Presently, however, only El Niño and La Niña, which last only a single winter, can be predicted in ocean circulation models with a good degree of accuracy up to a year in advance. Being able to predict long-term droughts is an area researchers are focusing on heavily for the future, especially because of the threat of global warming.

Seager has come up with some theories about the future of the American Southwest and other subtropical regions that will become more arid over time, especially because of global warming. This warming is not something that will occur in the future—it is happening right now and has been going on for several years. It will become an established condition in years to come and create a permanent drought condition. Seager sees this as one of the most critical issues for climate modelers to deal with. He states that 19 modeling groups around the world have looked at the anthropogenic causes of climate change and agree that regions in the subtropics, including the American Southwest, are headed toward much more arid climates in the future. According to the findings in the models, the effects of humans on the land and their overuse of resources and poor land management practices that result in desertification have become very pronounced in the models of the past few decades.

Seager also points out that unlike droughts of the past, models that predict the future do not show the outcome linked so much to natural phenomena as to the overall surface warming driven by increasing greenhouse gases introduced by humans—global warming. The main reason this result is so evident in all the models is that they do not rely on hard-to-model inputs, such as cloud physics, but instead are based on large-scale atmospheric dynamics, a straightforward component that generates nonrefutable output.

These results should be of great concern to many people. Continued drying of already arid lands in the southwestern United States and northern Mexico will have dire consequences for issues such as water resources, land development, economics, industry, political relations, travel, and migration in the years to come.

PANGAEA ULTIMA—THE FUTURE?

Scientists are interested in what the future may be like and how the Earth may eventually look. What will its climate be like? How will its ecosystems function? How many of the changes will be natural, and how many will be human caused? In the case of plate tectonics, geologists have been able to trace the movement of the Earth's plates backward in time in order to determine the geographic positions of the continents and the resulting climates that affected them, such as tropical, polar, or temperate.

The Earth's plate tectonic process continues today, and scientists wonder what the geographic configuration of the continents will one day become. Although these projections are theoretical, they present an intriguing view of what may lie in the future and what role climate may play.

Dr. Christopher R. Scotese, a geologist at the University of Texas, believes that projecting the locations of continents 50 million years from now is not difficult. Projecting beyond that, however, is much more problematic because unpredictable cause-and-effect incidents can drastically change results. "Fifty million years is fairly straightforward. It's like you're driving on the highway and you want to know where you're going to be in 10 minutes. You check the speedometer, do a calculation, and project your present motion. But beyond 50 million years—like on the highway, unexpected things can happen."

Projecting further into the future is more difficult because it involves far more than simple extrapolations. Instead, rules must be developed that govern not only their movements but also where subduction zones and deep ocean trenches will form. They may change shape but seldom disappear altogether because bedrock weighs little compared to the dense ocean crust. Continents literally float, as do mountains. Once formed, they tend to persist and disappear only after ages of erosion wear them down. The difficult part is to predict the development of new subduction zones in the seafloor and to calculate how rapidly such zones will rearrange the continents. "It's hard

(opposite page) Dr. Christopher R. Scotese's Paleomap Project. (A) This is what the Earth's distribution of continents may look like in 50 million years from now. The Atlantic may widen, Africa will collide with Europe closing the Mediterranean, Australia will collide with Southeast Asia, and California will slide northward up the coast to Alaska. (B) This is Pangaea Ultima 250 million years from now. It will form as a result of the subduction of the ocean floor of the North and South Atlantic beneath eastern North America and South America. This supercontinent will have a small ocean basin trapped at its center. (Source: www.scotese.com)

to understand all the forces down there," Scotese says. "There's probably some input from the mantle. It probably has some say on which way the plates go."

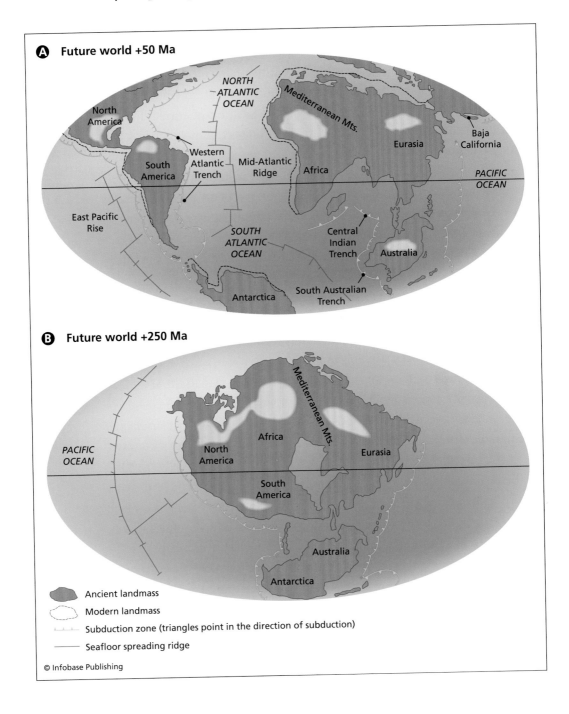

A Future world +50 Ma

NORTH ATLANTIC OCEAN

Mediterranean Mts.

North America

Baja California

Eurasia

Western Atlantic Trench

Mid-Atlantic Ridge

South America

Africa

PACIFIC OCEAN

East Pacific Rise

SOUTH ATLANTIC OCEAN

Central Indian Trench

Australia

South Australian Trench

Antarctica

B Future world +250 Ma

Mediterranean Mts.

PACIFIC OCEAN

Africa

North America

Eurasia

South America

Australia

Antarctica

Ancient landmass

Modern landmass

Subduction zone (triangles point in the direction of subduction)

Seafloor spreading ridge

Several geologists have predicted that in 50 million years, the tectonic movement of the San Andreas fault will have moved Los Angeles northward to the area of San Francisco. Eventually, Los Angeles will move north as far as Anchorage, Alaska.

In 1998, Dr. Scotese created a project called the Paleomap Project (www.scotese.com), which projects what the continents may look like 50 million years from now and 250 million years from now. He calls the configuration predicted for 250 million years from now Pangaea Ultima. In order to figure out the eventual location of continents, Dr. Scotese used a computer algorithm to simulate the mechanisms that move the plates, such as where subduction zones recycle and melt the plates, where trenches in the ocean floor tear continents apart, where ridges in the ocean floor tear the ocean floor apart and move continents away from one another, and where other continents are being pushed together, creating new mountain ranges.

Dr. Scotese believes the most difficult part of modeling the future distribution of the continents is predicting the development of new subduction zones and determining how aggressive they will be in moving the plates compared to older subduction zones. In addition to the forces on the crust, forces deep inside the Earth's mantle play a part in plate tectonics and are difficult to model.

Dr. Scotese predicts that the processes that will lead to the Earth 250 million years from now, Pangaea Ultima, will begin with the closing of the Mediterranean. Then, 25 to 75 million years from now, Australia will migrate north, collide with Indonesia and Malaysia, then turn counter-clockwise and collide with the Philippines and Asia, eventually merging them all together. In addition, Antarctica will migrate northward, upon which its icecap will melt. About 100 million years from now, it will enter the Indian Ocean. Then, 50 million years later, it will settle between Madagascar and Indonesia, making the Indian Ocean an inland sea.

The most drastic change by far will be the closing of the Atlantic Ocean. Then, 200 million years from now, Newfoundland will collide with Africa, and Brazil will butt up against South Africa. Finally, 250 million years from now, all the continents will have merged into a new super continent, Pangaea Ultima, that will encircle the remnants of the old Indian Ocean.

This is just one prediction of what the Earth may look like in the future. Other scientists have projected that instead of the Atlantic Ocean disappearing, the Pacific Ocean may disappear instead, pushing both North and South America into Asia and forming a hypothetical new continent called Amasia.

These are nevertheless hypothetical projections of the Earth's future created by computer models. While no one knows for sure what the Earth will look like in the distant future, it does bring up interesting questions about what the Earth's climate may be like. If the bulk of the continents are located near the equator, what will have happened to those ecosystems that were not tropical? Likewise, what about tropical habitats that now may reside in the mid-latitudes? In addition, if Antarctica migrates north and its ice cap completely melts, what will sea levels be? What will be the composition of seawater, and what will be the new configuration of the major ocean currents?

With a new spatial arrangement of the continents, there will undoubtedly be a new distribution of ocean currents, which in turn will affect global heat distribution. This will affect vegetation distributions, which, in turn, will affect the carbon cycle. Finally, in light of all these hypothetical changes and their effects on energy, heat, and the carbon cycle, what role will global warming have had on all this? Will global warming have been controlled 250 million years earlier so that there are still productive civilizations left to see what the actual Pangaea Ultima will look like? These are the questions climate scientists are working diligently to answer today.

THE GEOLOGICAL TIMESCALE

ERA	PERIOD		EPOCH	AGE (MILLIONS OF YEARS)	FIRST LIFE-FORMS	GEOLOGY
Cenozoic	Quaternary		Holocene	0.01	Humans	Ice age
			Pleistocene	3		
	Tertiary	Neogene	Pliocene	11	Mastodons	Cascades
			Miocene	26	Saber-toothed tigers	Alps
		Paleogene	Oligocene	37		
			Eocene	54	Whales	
			Paleocene	65	Horses, Alligators	Rockies
Mesozoic	Cretaceous			135	Birds Mammals Dinosaurs	Sierra Nevada Atlantic
	Jurassic			210		
	Triassic			250		
Paleozoic	Permian			280	Reptiles	Appalachians
	Carboniferous		Pennsylvanian	310	Trees	Ice age
			Mississippian	345	Amphibians Insects	Pangaea
	Devonian			400	Sharks	
	Silurian			435	Land plants	Laursia
	Ordovician			500	Fish	
	Cambrian			544	Sea plants Shelled animals	Gondwana
Proterozoic				700	Invertebrates	
				2500	Metazoans	
				3500	Earliest life	
Archean				4000		Oldest rocks
				4600		Meteorites

THE PERIODIC TABLE OF THE ELEMENTS

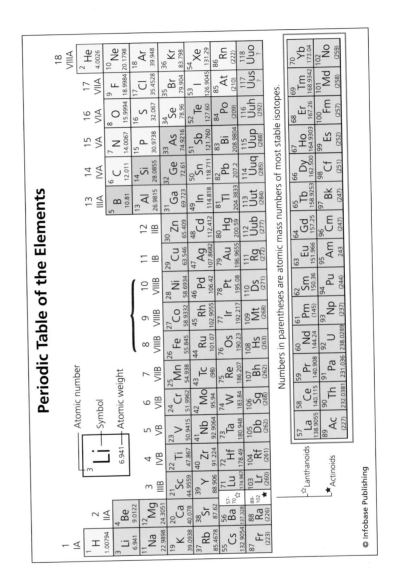

Periodic Table of the Elements

Atomic number
Symbol
Atomic weight

3
Li
6.941

Numbers in parentheses are atomic mass numbers of most stable isotopes.

☆ Lanthanoids
★ Actinoids

© Infobase Publishing

The Chemical Elements

(g) none (c) nonmetallics

element	symbol	a.n.
carbon	C	6
hydrogen	H	1

(g) chalcogen (c) nonmetallics

element	symbol	a.n.
oxygen	O	8
polonium	Po	84
selenium	Se	34
sulfur	S	16
tellurium	Te	52
ununhexium	Uuh	116

(g) alkali metal (c) metallics

element	symbol	a.n.
cesium	Cs	55
francium	Fr	87
lithium	Li	3
potassium	K	19
rubidium	Rb	37
sodium	Na	11

(g) alkaline earth metal (c) metallics

element	symbol	a.n.
barium	Ba	56
beryllium	Be	4
calcium	Ca	20
magnesium	Mg	12
radium	Ra	88
strontium	Sr	38

(g) none (c) metallics

element	symbol	a.n.
aluminum	Al	13
bohrium	Bh	107
cadmium	Cd	48
chromium	Cr	24
cobalt	Co	27
copper	Cu***	29
darmstadtium	Ds	110
dubnium	Db	105
gallium	Ga	31
gold	Au***	79
hafnium	Hf	72
hassium	Hs	108
indium	In	49
iridium	Ir****	77
iron	Fe	26
lawrencium	Lr	103
lead	Pb	82
lutetium	Lu	71
manganese	Mn	25
meitnerium	Mt	109
mercury	Hg	80
molybdenum	Mo	42
nickel	Ni	28
niobium	Nb	41
osmium	Os****	76
palladium	Pd****	46
platinum	Pt****	78
rhenium	Re	75
rodium	Rh****	45
roentgenium	Rg	111
ruthenium	Ru****	44
rutherfordium	Rf	104
scandium	Sc	21
seaborgium	Sg	106
silver	Ag***	47
tantalum	Ta	73
technetium	Tc	43
thallium	Tl	81
titanium	Ti	22
tin	Sn	50
tungsten	W	74
ununbium	Uub	112
ununtrium	Uut	113
ununquadium	Uuq	114
vanadium	V	23
yttrium	Y	39
zinc	Zn	30
zirconium	Zr	40

(g) pnictogen (c) metallics

element	symbol	a.n.
arsenic	As*	33
antimony	Sb*	51
bismuth	Bi	83
nitrogen	N	7
phosphorus	P**	15
ununpentium	Uup	115

(g) none (c) semimetallics

element	symbol	a.n.
boron	B	5
germanium	Ge	32
silicon	Si	14

(g) actinoid (c) metallics

element	symbol	a.n.
actinium	Ac	89
americium	Am	95
berkelium	Bk	97
californium	Cf	98
curium	Cm	96
einsteinium	Es	99
fermium	Fm	100
mendelevium	Md	101
neptunium	Np	93
nobelium	No	102
plutonium	Pu	94
protactinium	Pa	91
thorium	Th	90
uranium	U	92

(g) halogens (c) nonmetallics

element	symbol	a.n.
astatine	At*	85
bromine	Br	35
chlorine	Cl	17
fluorine	F	9
iodine	I	53
ununseptium	Uus*	117

(g) lanthanoid (c) metallics

element	symbol	a.n.
cerium	Ce	58
dysprosium	Dy	66
erbium	Er	68
europium	Eu	63
gadolinium	Gd	64
holmium	Ho	67
lanthanum	La	57
neodymium	Nd	60
praseodymium	Pr	59
promethium	Pm	61
samarium	Sm	62
terbium	Tb	65
thulium	Tm	69
ytterbium	Yb	70

(g) noble gases (c) nonmetallics

element	symbol	a.n.
argon	Ar	18
helium	He	2
krypton	Kr	36
neon	Ne	10
radon	Rn	86
xenon	Xe	54
ununoctium	Uuo	118

* = semimetallics (c)
** = nonmetallics (c)
*** = coinage metal (g)
**** = precious metal (g)

a.n. = atomic number
(g) = group
(c) = classification

© Infobase Publishing

CHRONOLOGY

ca. 1400–1850   Little Ice Age covers the Earth with record cold, large glaciers, and snow. There is widespread disease, starvation, and death.

1800–70   The levels of CO_2 in the atmosphere are 290 ppm.

1824   Jean-Baptiste Joseph Fourier, a French mathematician and physicist, calculates that the Earth would be much colder without its protective atmosphere.

1827   Jean-Baptiste Joseph Fourier presents his theory about the Earth's warming. At this time many believe warming is a positive thing.

1859   John Tyndall, an Irish physicist, discovers that some gases exist in the atmosphere that block infrared radiation. He presents the concept that changes in the concentration of atmospheric gases could cause the climate to change.

1894   Beginning of the industrial pollution of the environment.

1913–14   Svante Arrhenius discovers the greenhouse effect and predicts that the Earth's atmosphere will continue to warm. He predicts that the atmosphere will not reach dangerous levels for thousands of years, so his theory is not received with any urgency.

1920–25   Texas and the Persian Gulf bring productive oil wells into operation, which begins the world's dependency on a relatively inexpensive form of energy.

1934   The worst dust storm of the dust bowl occurs in the United States on what historians would later call Black Sunday. Dust storms are a product of drought and soil erosion.

1945   The U.S. Office of Naval Research begins supporting many fields of science, including those that deal with climate change issues.

1949–50   Guy S. Callendar, a British steam engineer and inventor, propounds the theory that the greenhouse effect is linked to human actions and will cause problems. No one takes him too seriously, but scientists do begin to develop new ways to measure climate.

1950–70   Technological developments enable increased awareness about global warming and the enhanced greenhouse effect. Studies confirm a steadily rising CO_2 level. The public begins to notice and becomes concerned with air pollution issues.

1958   U.S. scientist Charles David Keeling of the Scripps Institution of Oceanography detects a yearly rise in atmospheric CO_2. He begins collecting continuous CO_2 readings at an observatory on Mauna Loa, Hawaii. The results became known as the famous Keeling Curve.

1963   Studies show that water vapor plays a significant part in making the climate sensitive to changes in CO_2 levels.

1968   Studies reveal the potential collapse of the Antarctic ice sheet, which would raise sea levels to dangerous heights, causing damage to places worldwide.

1972   Studies with ice cores reveal large climate shifts in the past.

1974   Significant drought and other unusual weather phenomenon over the past two years cause increased concern about climate change not only among scientists but with the public as a whole.

1976   Deforestation and other impacts on the ecosystem start to receive attention as major issues in the future of the world's climate.

1977   The scientific community begins focusing on global warming as a serious threat needing to be addressed within the next century.

1979   The World Climate Research Programme is launched to coordinate international research on global warming and climate change.

1982   Greenland ice cores show significant temperature oscillations over the past century.

1983   The greenhouse effect and related issues get pushed into the political arena through reports from the U.S. National Academy of Sciences and the Environmental Protection Agency.

1984–90   The media begins to make global warming and its enhanced greenhouse effect a common topic among Americans. Many critics emerge.

1987   An ice core from Antarctica analyzed by French and Russian scientists reveals an extremely close correlation between CO_2 and temperature going back more than 100,000 years.

1988   The United Nations set up a scientific authority to review the evidence on global warming. It is called the Inter-governmental Panel on Climate Change (IPCC) and consists of 2,500 scientists from countries around the world.

1989   The first IPCC report says that levels of human-made greenhouse gases are steadily increasing in the atmosphere and predicts that they will cause global warming.

1990   An appeal signed by 49 Nobel prizewinners and 700 members of the National Academy of Sciences states, "There is broad agreement within the scientific community that amplification of the Earth's natural greenhouse effect by the buildup of various gases introduced by human activity has the potential to produce dramatic changes in climate . . . Only by taking action now can we insure that future generations will not be put at risk."

1992   The United Nations Conference on Environment and Development (UNCED), known informally as the Earth Summit, begins on June 3 in Rio de Janeiro, Brazil. It results in the United Nations Framework Convention on Climate Change, Agenda 21, the Rio Declaration on Environment and Development Statement of Forest Principles, and the United Nations Convention on Biological Diversity.

1993   Greenland ice cores suggest that significant climate change can occur within one decade.

1995   The second IPCC report is issued and concludes there is a human-caused component to the greenhouse effect warming. The consensus is that serious warming is likely in the coming century. Reports on the breaking up of Antarctic ice sheets and other signs of warming in the polar regions are now beginning to catch the public's attention.

1997   The third conference of the parties to the Framework Convention on Climate Change is held in Kyoto, Japan. Adopted on December 11, a document called the Kyoto Protocol commits its signatories to reduce emissions of greenhouse gases.

2000   Climatologists label the 1990s the hottest decade on record.

2001   The IPPC's third report states that the evidence for anthropogenic global warming is incontrovertible, but that its effects on climate are still difficult to pin down. President Bush declares scientific uncertainty too great to justify Kyoto Protocol's targets.

The United States Global Change Research Program releases the findings of the National Assessment of the Potential Consequences of Climate Variability and Change. The assessment finds that temperatures in the United States will rise by 5 to 9°F (3–5°C) over the next century and predicts increases in both very wet (flooding) and very dry (drought) conditions. Many ecosystems are vulnerable to climate change. Water supply for human consumption and irrigation is at risk due to increased probability of drought, reduced snow pack, and increased risk of flooding. Sea-level rise and storm surges will most likely damage coastal infrastructure.

2002   Second hottest year on record.

Heavy rains cause disastrous flooding in Central Europe leading to more than 100 deaths and more than $30 billion in damage. Extreme drought in many parts of the world (Africa, India,

Australia, and the United States) results in thousands of deaths and significant crop damage. President Bush calls for 10 more years of research on climate change to clear up remaining uncertainties and proposes only voluntary measures to mitigate climate change until 2012.

2003 U.S. senators John McCain and Joseph Lieberman introduce a bipartisan bill to reduce emissions of greenhouse gases nationwide via a greenhouse gas emission cap and trade program.

Scientific observations raise concern that the collapse of ice sheets in Antarctica and Greenland can raise sea levels faster than previously thought.

A deadly summer heat wave in Europe convinces many in Europe of the urgency of controlling global warming but does not equally capture the attention of those living in the United States.

International Energy Agency (IEA) identifies China as the world's second largest carbon emitter because of their increased use of fossil fuels.

The level of CO_2 in the atmosphere reaches 382 ppm.

2004 Books and movies feature global warming.

2005 Kyoto Protocol takes effect on February 16. In addition, global warming is a topic at the G8 summit in Gleneagles, Scotland, where country leaders in attendance recognize climate change as a serious, long-term challenge.

Hurricane Katrina forces the U.S. public to face the issue of global warming.

2006 Former U.S. vice president Al Gore's *An Inconvenient Truth* draws attention to global warming in the United States.

Sir Nicholas Stern, former World Bank economist, reports that global warming will cost up to 20 percent of worldwide gross domestic product if nothing is done about it now.

2007 IPCC's fourth assessment report says glacial shrinkage, ice loss, and permafrost retreat are all signs that climate change is underway now. They predict a higher risk of drought, floods, and more powerful storms during the next 100 years. As a result, hunger, homelessness, and disease will increase. The atmosphere may warm 1.8 to 4.0°C and sea levels may rise 7 to 23 inches (18 to 59 cm) by the year 2100.

Al Gore and the IPCC share the Nobel Peace Prize for their efforts to bring the critical issues of global warming to the world's attention.

2008   The price of oil reached and surpassed $100 per barrel, leaving some countries paying more than $10 per gallon.

Energy Star appliance sales have nearly doubled. Energy Star is a U.S. government-backed program helping businesses and individuals protect the environment through superior energy efficiency.

U.S. wind energy capacity reaches 10,000 megawatts, which is enough to power 2.5 million homes.

2009   President Obama takes office and vows to address the issue of global warming and climate change by allowing individual states to move forward in controlling greenhouse gas emissions. As a result, American automakers can prepare for the future and build cars of tomorrow and reduce the country's dependence on foreign oil. Perhaps these measures will help restore national security and the health of the planet, and the U.S. government will no longer ignore the scientific facts.

The year 2009 will be a crucial year in the effort to address climate change. The meeting on December 7–18 in Copenhagen, Denmark, of the UN Climate Change Conference promises to shape an effective response to climate change. The snapping of an ice bridge in April 2009 linking the Wilkins Ice Shelf (the size of Jamaica) to Antarctic islands could cause the ice shelf to break away, the latest indication that there is no time to lose in addressing global warming.

GLOSSARY

adaptation   an adjustment in natural or human systems to a new or changing environment. Adaptation to climate change refers to adjustments in natural or human systems in response to actual or expected climatic changes.

aerosols   tiny bits of liquid or solid matter suspended in air. They come from natural sources such as erupting volcanoes and from waste gases emitted from automobiles, factories, and power plants. By reflecting sunlight, aerosols cool the climate and offset some of the warming caused by greenhouse gases.

albedo   the relative reflectivity of a surface. A surface with high albedo reflects most of the light that shines on it and absorbs very little energy; a surface with low albedo absorbs most of the light energy that shines on it and reflects very little.

anemometer   a device for measuring wind speed.

anthropogenic   made by people or resulting from human activities. This term is usually used in the context of emissions produced as a result of human activities.

atmosphere   the thin layer of gases that surrounds the Earth and allows living organisms to breathe. It reaches 400 miles (644 km) above the surface, but 80 percent is concentrated in the troposphere—the lower seven miles (11 km) above the Earth's surface.

barometer   an instrument that measures atmospheric pressure.

biodiversity   different plant and animal species.

biomass   plant material that can be used for fuel.

bleaching (coral)   the loss of algae from corals that causes the corals to turn white. This is one of the results of global warming and signifies a die-off of unhealthy coral.

carbon dioxide   a colorless, odorless gas that forms when carbon atoms combine with oxygen atoms. Carbon dioxide is a tiny but vital part of the atmosphere. The heat-absorbing ability of carbon dioxide is what makes life possible on Earth.

carbon sink   an area where large quantities of carbon are built up in the wood of trees, in calcium carbonate rocks, in animal species, in the oceans and in any other place where carbon is stored. These places act as reservoirs, keeping carbon out of the atmosphere.

chlorofluorocarbons (CFCs)   gases that were once widely used as coolants in refrigerators and air conditioners, as foaming agents for insulation and food packaging, and as cleaning agents in certain industries. They are long-lasting compounds that absorb heat energy more effectively than carbon dioxide. When they enter the upper atmosphere, they destroy ozone, which protects life on Earth from harmful ultraviolet radiation. An international treaty calls for all production of CFCs to stop by 2010.

climate   the usual pattern of weather averaged over a long period of time.

climate feedback   An interaction mechanism between processes in the climate system whereby the result of an initial process triggers changes in a second process that in turn influences the initial one. A positive feedback intensifies the original process, and a negative feedback reduces it.

climate model   a quantitative way of representing the interactions of the atmosphere, oceans, land surface, and ice. Models can range from relatively simple to extremely complicated.

climatologist   a scientist who studies the climate.

concentration   the amount of a component in a given area or volume. In global warming, it is a measurement of how much of a particular gas is in the atmosphere compared to all the gases in the atmosphere.

condensation   the process that changes a gas into a liquid.

deforestation   the large-scale cutting of trees from a forested area, often leaving large areas bare and susceptible to erosion.

ecosystem   a community of interacting organisms and their physical environment.

El Niño   a cyclic weather event in which the waters of the eastern Pacific Ocean off the coast of South America become much warmer than normal and disturb weather patterns across the region. Its full name is El Niño Southern Oscillation (ENSO). Every few years, the temperature of the western Pacific rises several degrees above that of waters to the east. The warmer water moves eastward, causing shifts in ocean currents, jet stream winds, and weather in both the Northern and Southern Hemispheres.

emissions   the release of a substance (usually a gas when referring to climate change) into the atmosphere.

evaporation   the process by which a liquid, such as water, is changed to a gas.

evapotranspiration   the transfer of moisture from the Earth to the atmosphere by evaporation of water and transpiration from plants.

feedback   a change caused by a process that in turn may influence that process. Some changes caused by global warming may hasten the process of warming (positive feedback); some may slow warming (negative feedback).

fossil fuel   an energy source made from coal, oil, or natural gas. The burning of fossil fuels is one of the chief causes of global warming.

glacier   a mass of ice formed by the build-up of snow over hundreds and thousands of years.

global warming   an increase in the temperature of the Earth's atmosphere caused by the build-up of greenhouse gases; also referred to as the enhanced greenhouse effect caused by humans.

global warming potential (GWP)   Global warming potential (GWP) is the cumulative radiative forcing effects of a gas over a specified time resulting from the emission of a unit mass of gas relative to a reference gas (usually CO_2).

greenhouse effect   the natural trapping of heat energy by gases present in the atmosphere, such as carbon dioxide, methane, and water vapor. The trapped heat is then emitted as heat back to the Earth.

greenhouse gas   a gas that traps heat in the atmosphere and keeps the Earth warm enough to allow life to exist.

Gulf Stream   a warm current that flows from the Gulf of Mexico across the Atlantic Ocean to northern Europe. It is largely responsible for Europe's mild climate.

humidity, relative   the amount of water vapor in the air expressed as a percentage of the maximum amount the air could hold at that given temperature.

industrial revolution   the period during which industry developed rapidly as a result of advances in technology. This took place in Britain during the late 18th and early 19th centuries.

infrared   the invisible heat radiation emitted by the Sun and by virtually every warm substance or object on Earth.

Intergovernmental Panel on Climate Change (IPCC)   an organization consisting of 2,500 scientists that assesses information in the scientific and technical literature related to the issue of climate change. The IPCC was established jointly by the United Nations Environment Programme and the World Meteorological Organization in 1988.

land use   the management practice of a certain land cover type. Land use may be such things as forest, arable land, grassland, urban land, and wilderness.

land use change   an alteration of the management practice on a certain land cover type. Land use changes may influence climate systems because they affect evapotranspiration and sources and sinks of greenhouse gases. An example of land use change is removing a forest to build a city.

methane   a colorless, odorless, flammable gas that is the major ingredient of natural gas. Methane is produced wherever decay occurs and little or no oxygen is present.

monsoon   heavy rains that occur at the same time each year.

nitrogen   as a gas, nitrogen makes up 80 percent of the volume of the Earth's atmosphere. It is also an element in substances such as fertilizer.

nitrous oxide   a heat-absorbing gas in the Earth's atmosphere. Nitrous oxide is emitted from nitrogen-based fertilizers.

nuclear power   the electricity produced by a process that begins with the splitting apart of uranium atoms, yielding great amounts of heat energy.

ozone   a molecule that consists of three oxygen atoms. Ozone is present in small amounts in the Earth's atmosphere at 14 to 19 miles (23–31 km) above the Earth's surface. A layer of ozone makes life possible by shielding the Earth's surface from most harmful ultraviolet rays. In the lower atmosphere, ozone emitted from auto exhausts and factories is an air pollutant.

parts per million (ppm)   the number of parts of a chemical found in one million parts of a particular gas, liquid, or solid.

permafrost   permanently frozen ground in the Arctic. As global warming increases, this ground is melting.

photosynthesis   the process by which plants make food using light energy, carbon dioxide, and water.

protocol   the terms of a treaty that have been agreed to and signed by all parties.

radiation   the particles or waves of energy.

rain gauge   an instrument for measuring rainfall.

relative humidity   *see* humidity, relative.

renewable   something that can be replaced or regrown, such as trees, or a source of energy that never runs out, such as solar energy, wind energy, and geothermal energy.

resources   the raw materials from the Earth that are used by humans to make useful things.

satellite   any small object that orbits a larger one. Artificial satellites carry instruments for scientific study and communication. Imagery taken from satellites is used to monitor aspects of global warming such as glacier retreat, ice cap melting, desertification, erosion, hurricane damage, and flooding. Sea surface temperatures and measurements are also obtained from human-made satellites in orbit around the Earth.

simulation   a computer model of a process based on actual facts. The model attempts to mimic, or replicate, actual physical processes.

temperate   an area that has a mild climate and different seasons.

thermal   related to heat.

tropical   a region that is hot and often wet (humid). These areas are located around the Earth's equator.

tundra   a vast treeless plain in the Arctic with a marshy surface covering a permafrost layer.

ultraviolet radiation   a portion of the Sun's electromagnetic spectrum that consists of very short wavelengths and high energy. The atmosphere's ozone layer protects life on Earth from the damaging effects of UV radiation.

weather   the conditions of the atmosphere at a particular time and place. Weather includes such measurements as temperature, precipitation, air pressure, and wind speed and direction.

FURTHER RESOURCES

PRINT
Books

Christianson, Gale. *Greenhouse: The 200-Year Story of Global Warming.* New York: Walker, 1999. This book looks at the enhanced greenhouse effect worldwide after the Industrial Revolution and outlines the consequences to the environment.

Friedman, Katherine. *What if the Polar Ice Caps Melted?* Danbury, Conn.: Children's Press, 2002. This book focuses on environmental problems related to the Earth's atmosphere, including global warming, changing weather patterns, and effects on ecosystems.

Gelbspan, Ross. *The Heat Is On: The High Stakes Battle over Earth's Threatened Climate.* Reading, Mass.: Addison Wesley, 1997. This work offers a look at the controversy environmentalists often face when they deal with fossil fuel companies.

Harrison, Patrick "GB," Gail "Bunny" McLeod, and Patrick G. Harrison. *Who Says Kids Can't Fight Global Warming.* Chattanooga, Tenn.: Pat's Top Products, 2007. This book offers real steps that everybody can take to help solve the world's biggest air pollution problems.

Houghton, John. *Global Warming: The Complete Briefing.* New York: Cambridge University Press, 2004. This book outlines the scientific basis of global warming and describes the effects that climate change will have on society. It also looks at solutions to the problem.

Langholz, Jeffrey. *You Can Prevent Global Warming (and Save Money!): 51 Easy Ways.* Riverside, N.J.: Andrews McMeel Publishing, 2003. This book aims to convert public concern over global warming into positive action to stop it by providing simple, everyday practices that can easily be done to minimize it as well as save money.

McKibben, Bill. *Fight Global Warming Now: The Handbook for Taking Action in Your Community.* New York: Holt Paperbacks, 2007. This book provides the facts of what must change to save the climate. It also shows how everyone can act proactively in the community to make a difference.

Pringle, Laurence. *Global Warming: The Threat of Earth's Changing Climate.* New York: SeaStar Publishing, 2001. This book provides information on the carbon cycle, rising sea levels, El Niño, aerosols, smog, flooding, and other issues related to global warming.

Thornhill, Jan. *This Is My Planet—The Kids Guide to Global Warming.* Toronto: Maple Tree Press, 2007. This book offers students the tools they need to become ecologically oriented by taking a comprehensive look at climate change in polar, ocean, and land-based ecosystems.

Weart, Spencer R. *The Discovery of Global Warming (New Histories of Science, Technology, and Medicine).* Cambridge, Mass.: Harvard University Press, 2004. This book traces the history of the global warming concept through a long process of incremental research rather than a dramatic revelation.

Journal Articles

Brahic, Catherine. "Ancient Mega-lake Discovered in Darfur." *New-Scientist* (4/12/07). Available online. URL: www.newscientist.com/article/dn11593-ancient-megalake-discovered-in-darfur.html. Accessed January 4, 2009. This article discusses how remote sensing was used to locate an immense underground reservoir in the Middle East.

Broad, William J. "Long-Term Global Forecast? Fewer Continents." *New York Times* (1/9/07). Available online. URL: www.nytimes.com/2007/01/09/science/09geo.html?r=1&oref=slogin. Accessed October 9, 2008. This article presents ideas on where the Earth's continents will one day be and how climate will affect them in the future.

Culotta, Elizabeth. "Will Plants Profit From High CO_2?" *Science* (5/5/95). This article explores the possible effects of various carbon dioxide levels on vegetation as a result of global warming and whether they will experience enhanced growth.

D'Agnese, Joseph. "Why Has Our Weather Gone Wild?" *Discover* (June 2000). This article focuses on the recent global changes in weather, such as shifting seasons, severe storms, droughts, heat waves, and other weather-related events and their connection to global warming.

Hoffman, Paul F., and Daniel P. Schrag. "Snowball Earth." *Scientific American* (January 2000). This article presents evidence that global climate change put the Earth in the grip of a completely frozen state eons ago.

Karl, Thomas, and Kevin Trenberth. "The Human Impact on Climate." *Scientific American* (December 1999). This article focuses on the disruptions people cause in the natural environment and why scientists must begin to monitor and quantify the disruptions now in order to save the future.

LeGrande, Allegra, and Gavin Schmidt. "Modeling an Abrupt Climate Change." *NASA GISS Science Briefs* (January 2006). Available online. URL: http://www.giss.nasa.gov/research/briefs/legrande_01/. Accessed October 9, 2008. This article discusses the objectives of modeling climate and why past events relate to the future.

Lovgren, Stefan. "Climate Change Killed Off Maya Civilization, Study Says." National Geographic News (3/13/03). Available online. URL: http://news.nationalgeographic.com/news/2003/03/0313_030313_mayadrought.html. Accessed October 9, 2008. This article looks at the influence climate had on ancient cultures.

Phillips, Mark. "Monumental Climate Change." CBS News London (6/19/07). Available online. URL: http://www.cbsnews.com/stories/2007/06/19/eveningnews/main2952286.shtml?source=related_story. Accessed October 9, 2008. This article provides information about how climate change is affecting many of the world's man-made monuments.

ScienceDaily. *Pack Rat Middens Give Unique View on Evolution and Climate Change in Past Million Years* (10/31/03). Available online.

URL: http://www.sciencedaily.com/releases/2003/10/031031062755. htm. Accessed October 9, 2008. This article discusses how pack rat middens enable paleoclimatologists to look at specific biologic records in order to understand climates of the ancient past.

Smol, John P., and Marianne S. V. Douglas. "Crossing the Final Ecological Threshold in High Arctic Ponds." *Proceedings of the National Academy of Sciences of the United States of America* (7/2/07). Available online. URL: http://www.pnas.org/content/104/30/12395. abstract. Accessed October 9, 2008. This article discusses biodiversity in arctic environments.

Time, various editors. *Global Warming: The Causes, The Perils, the Politics—and What It Means for You.* New York: *Time* (2007). This specially bound volume contains an extensive article that suggests 51 ways to save the environment and curb global warming.

WEB SITES

Global Warming

Climate Ark homepage. Ecological Internet. Available online. URL: www.climateark.org. Accessed October 23, 2007. A Web site that promotes public policy that addresses global climate change through reduction in carbon and other emissions, energy conservation, alternative energy sources, and ending deforestation.

Climate Solutions homepage. Atmosphere Alliance and Energy Outreach Center. Available online. URL: www.climatesolutions.org. Accessed October 23, 2007. A Web site that offers practical solutions to global warming.

Environmental Defense Fund homepage. Environmental Defense Fund. Available online. URL: www.environmentaldefense.org. Accessed October 26, 2007. A Web site from an organization started by a handful of environmental scientists in 1967 that provides quality information and helpful resources on understanding global warming and other crucial environmental issues.

Environmental Protection Agency homepage. U.S. Environmental Protection Agency. Available online. URL: www.epa.gov. Accessed

October 26, 2007. This Web site provides information about the EPA's efforts and programs to protect the environment. It offers a wide array of information on global warming.

European Environment Agency homepage. European Environment Agency in Copenhagen, Denmark. Available online. URL: www.eea. europa.eu/themes/climate. Accessed October 26, 2007. This Web site posts reports on topics such as air quality, ozone depletion, and climate change.

Global Warming: Focus on the Future homepage. EnviroLink. Available online. URL: www.enviroweb.org. Accessed October 26, 2007. This Web site offers statistics and photography of global warming topics.

HotEarth.net homepage. National Environmental Trust. Available online. URL: www.net.org/warming. Accessed October 26, 2007. This Web site features informational articles on the causes of global warming, its harmful effects, and solutions that could stop it.

Intergovernmental Panel on Climate Change (IPCC) homepage. World Meteorological Organization (WMO) and the United Nations Environment Programme (UNEP). Available online. URL: http://www.ipcc.ch/. Accessed October 26, 2007. This Web site offers current information on the science of global warming and recommendations on practical solutions and policy management.

NASA Goddard Institute for Space Studies homepage. National Aeronautics and Space Administration. Available online. URL: www.giss.nasa.gov. Accessed October 26, 2007. This Web site provides a large database of information, research, and other resources.

NOAA National Climatic Data Center homepage. National Oceanic and Atmospheric Administration. Available online. URL: www. ncdc.noaa.gov. Accessed October 26, 2007. This Web site offers a multitude of resources and information on climate, climate change, and global warming.

Ozone Action homepage. Southeast Michigan Council of Governments. Available online. URL: www.semcog.org/services/ozoneaction/kids.htm. Accessed October 26, 2007. This Web site provides information on air quality by focusing on ozone, the atmosphere, environmental issues, and related health issues.

Scientific American homepage. Scientific American, Inc. Available online. URL: www.sciam.com. Accessed October 23, 2007. This organization offers an online magazine and often presents articles concerning climate change and global warming.

Tyndall Centre at University of East Anglia homepage. Tyndall Centre for Climate Change Research. Available online. URL: http://www.tyndall.ac.uk. Accessed October 26, 2007. This Web site offers information on climate change and is considered one of the leaders in research on global warming.

Union of Concerned Scientists homepage. Union of Concerned Scientists. Available online. URL: www.ucsusa.org. Accessed October 26, 2007. This Web site offers quality resource sections on global warming and ozone depletion.

United Nations Framework Convention on Climate Change (UNFCCC) homepage. United Nations Framework Convention on Climate Change. Available online. URL: http://unfccc.int/2860.php. Accessed October 26, 2007. This Web site presents a wide array of climate change information and policy.

U.S. Global Change Research Program homepage. U.S. Office of Science and Technology Policy, Office of Management and Budget, and Council on Environmental Quality. Available online. URL: www.usgcrp.gov. Accessed October 26, 2007. This Web site provides information on the current research activities of national and international science programs that focus on global monitoring of climate and ecosystems.

World Wildlife Foundation Climate Change Campaign homepage. World Wildlife Fund. Available online. URL: www.worldwildlife.org/climate/. Accessed October 26, 2007. This Web site contains information on what various countries are doing, and not doing, to deal with global warming.

Greenhouse Gas Emissions

Energy Information Administration homepage. U.S. Department of Energy. Available online. URL: www.eia.doe.gov/environment.html. Accessed October 26, 2007. This Web site lists official environmental energy-related emissions data and environmental analyses from the U.S. government. The site contains U.S. carbon dioxide, methane, and nitrous oxide emissions data and other greenhouse reports.

World Resources Institute—Climate, Energy & Transport homepage. World Resources Institute. Available online. URL: www.wri.org/climate/publications.cfm. Accessed October 26, 2007. This Web site offers a collection of reports on global technology deployment to stabilize emissions, agriculture, greenhouse gas mitigation, climate science discoveries, and renewable energy.

INDEX

Italic page numbers indicate illustrations or maps. Page numbers followed by *c* denote entries in the chronology.